Marine Geology

A Planet Earth Perspective

Revised Printing

Roger N. Anderson

Lamont-Doherty Geological Observatory

Columbia University

John Wiley & Sons

New York Chichester Brisbane Toronto Singapore

Cover photograph by Sunny Redmond

Library of Congress Cataloging in Publication Data:

Anderson, Roger N.
  Marine geology.

  Bibliography: p.
  Includes index.
  1. Submarine geology.   I. Title.

QE39.A53   1986   551.46′08   85-22685
ISBN 0-471-50407-6

Printed in the United States of America

10 9 8 7 6 5 4 3 2

Preface

For several years now, I have taught a course called 'Physics and Chemistry of the Earth' in Columbia University's Science Honors Program. Concurrently, I teach 'Planet Earth' to undergraduates at Columbia. Although the Science Honors and Columbia College students are often brilliant academically, they are also *dogmatic* and set in their ways. Among the common problems I have are convincing these young students that all the problems in science are not already solved; that all they read is not necessarily correct; and that a scientist does not recite facts during his or her professional career. Instead, scientists ask questions that do not yet have answers, and then determine the ways to answer these questions. For instance, students are often surprised to learn that scientists write for a living. In many ways, my course is a microcosm of why the founders of the Science Honors Program began this curriculum. We are charged with conveying how science works more than just teaching the facts of science. The smarter a student is, especially early in a career, the harder it is to teach him or her that science is asking questions that have not yet been answered, rather than repeating facts discovered by others.

This book grew out of both these courses as a technique for demonstrating how earth scientists determine how the Earth works by logic, deduction, and inference. A researcher doing active science thinks very differently than a student preparing for an undergraduate degree. Often the student does not grasp how science works until well into a graduate education, if then. Here, in this text I try to not only recite facts but to develop the method and technique for how the questions were asked in the first place, and how the scientists went about discovering the answers to how the Earth works. The subject provides an excellent training ground for the scientific technique because most of the inner workings of Planet Earth are inaccessible to direct observation. Remote sensing methods must be used, and nowhere more so than in the study of the geology under the sea.

This book is not just a marine geology textbook, but is an attempt to ask questions of how the Earth works from a marine geology perspective. Consequently, we begin under the oceans, but end on the continents. It is designed as an undergraduate text; as a first introduction of the student to how an earth scientist studies the Earth. The book can therefore be used as an earth sciences introductory text as well as a beginning marine geology book.

Several people contributed especially to the conception and form of this book: Wally Broecker, Allan Sachs, Ken Macdonald, Bob Ballard, Jim Kennett and his graduate text on Marine Geology, Press and Siever's "Earth," *The Sea*, Vol. 7, Wiley-Interscience, edited by Ceasare Emiliani, and *Hydrothermal Processes at Sea Floor Spreading Centers*, edited by Peter Rona and others. Special thanks go to John Dewey for teaching me how to approach an undergraduate classroom, and for many fine illustrations used in this book. I benefited from the invaluable assistance of Hester Cason.

My wife, Honor, gave special support and encouragement, without which this book would have never come to fruition.

More than anything, it is truly exciting to study the geology under the sea and beneath the continents. The earth sciences are fun and exciting professions. This book attempts to convey that excitement, and I hope students will have fun reading it.

Roger N. Anderson

September, 1988

Contents

Preface

The inspiration for this book came from numerous buildings, books and people. Two books that have remained with us since our student days — Rasmussen's *Experiencing Architecture* and Heschong's *Thermal Delight in Architecture* — conveyed to us a sense of enjoyment of architecture that went beyond the spatial and stylistic. This was reinforced by the Cambridge Masters course in Environmental Design in Architecture, which we both attended, founded and coordinated at that time by Nick Baker and Dean Hawkes. If there is an axiom that came out of that education then it could read as 'Building science is at the service of architecture, and architecture at the service of people'.

From the mid-1990s we joined forces to run a design studio at Cambridge that specifically combined the experiential with the environmental in architectural design projects. Simultaneously, a series of our doctoral and masters students at the Department of Architecture — many of whom are authors in this book (Fisher, Merghani, Nikolopoulou, Parpairi, Potvin, Ramos and Sinou) — began to explore the links between the measurable and the perceived environmental characteristics of architecture. Along the way we came across sympathetic research and consultancy colleagues who were in a position to offer valuable perspectives that have been included in this book. A body of work started taking shape, which provided the arguments and evidence underpinning the notions of 'environmental diversity in architecture'.

In 1988, Dean Hawkes identified the following challenge for the field of environmental research: 'the need to direct studies of user requirements towards the understanding of environmental diversity, both spatial and temporal, and of the complex perceptual and operational relationships which occur in the total environment' (Hawkes 1996: 105).* We hope this book goes some way to meeting that challenge.

Koen Steemers and Mary Anne Steane
Cambridge
July 2004

---

* Hawkes, D. (1996) *The Environmental Tradition*, London: Spon (originally published in Kroner, W. M. (ed.) (1988) 'A new frontier: environments for innovation', *Proceedings of the International Symposium on Advanced Comfort System for the Work-Environment, 1–3 May*, New York: Rensselaer Polytechnic Institute).

Acknowledgements

We would like to acknowledge first and foremost all the authors in this book for their contributions and patience. In particular we thank Nick Baker, who as tutor and colleague provided much of the inspiration for this book, and Peter Carl who contributed to many hours of discussion, and read and supported our efforts with indefatigable enthusiasm. Although not directly involved in this book, Dean Hawkes has given us tremendous motivation in his roles as academic and practitioner of environmentally responsive architecture, leading by example.

To Jeanette, Kai and Finn — KS.
To my parents — MAS.

Contributors

Nick Baker, BSc, MA, PhD

Nick studied physics at London University, going on to work in a Medical Research Council team on the possible genetic effects of ultrasound. However, for most of his professional life he has worked in the field of environmental building science, originally teaching at NE London Polytechnic (now UEL) and, since 1985, at the Martin Centre for Architectural and Urban Studies, University of Cambridge. During this time he has been involved with several national and European research projects on low-energy buildings, thermal comfort, daylighting and urban microclimate. He is the originator of the LT Method, an energy design tool, and the author and co-author of a number of books including *Low Energy and Passive Design in Tropical Island Climates* (1996), *Energy and Environment in Architecture* (2000) and *Daylighting Design in Buildings* (2002). He is currently technical expert on the EU-funded project REVIVAL, concerned with low energy refurbishment of large post-war buildings, as well as visiting lecturer to the MPhil Course in Environmental Design at Cambridge.

Peter Carl, BArch, MArch

Peter studied architecture at Princeton and has taught in the Department of Architecture at the University of Cambridge since 1979. Prior to that, he taught at the University of Kentucky and spent two years in Rome as a Prix de Rome scholar. At Cambridge, he is a convenor of the graduate programme in the History and Philosophy of Architecture, which seeks to understand how architectural and urban order contribute to culture as a whole. He is also a teacher in the graduate design programme, where much of the material regarding Sustainable Urban Metabolism has been developed in collaboration with Dalibor Vesely, Phillip Meadowcroft and Koen Steemers. His present research interest concerns transforming the benign ideal of 'mixed use' into a more coherent understanding of urban political topography.

John E. Fernandez, BSc, MArch, RA, AIA

John has been an Assistant Professor of Design and Building Technology in the School of Architecture and Planning at the Massachusetts Institute of Technology since 1999. His expertise and research addresses the field of architectural materials and building lifetimes. His teaching responsibilities include both technology and design courses taught to undergraduate and graduate students. He is currently engaged in several research projects investigating the use of innovative materials in constructional systems and is finishing a book, entitled *Emergent Materials*, which outlines the most important developments in materials for contemporary architecture. John has been awarded a teaching citation from the Graduate Student Council at MIT. John holds a Bachelor of Science degree from MIT in 1985 and a Master of Architecture degree from Princeton in 1989. Before entering the academic world John was a senior designer at both Kohn Pedersen Fox (1989–92) and Polshek and Partners (1993–97), both of New York City.

Peter Fisher, BA, BArch, MPhil, RIBA

Peter is an architect who works in London with Bennetts Associates. He studied architecture and environmental design in Hull and Cambridge and has worked for a number of years in both Germany and England, where key projects have included the Leipzig Trade Fair with von Gerkan Marg and the recently completed Sophos Headquarters in Abingdon with Bennetts. He has a strong interest in the history and evolution of the building envelope and much of his practise experience has been in the design and construction of cladding. His paper submitted to PLEA's 2000 Conference, on a low-energy speculative office design, was jointly awarded 'Best Short Paper' and his design for the RIBA Sustainable School Competition was short-listed in 2001. He has taught in Cardiff and Cambridge and recently sat on the BRE/M4I steering group for the construction traffic environmental performance indicator.

Tim Lewers, BSc, MSc, MIOA

Tim is an independent acoustic consultant. Recent buildings for which he has offered advice include The Cambridge Arts Theatre, The Contact Theatre in Manchester and The Garrick in Lichfield. The problems of speech reinforcement in reverberant spaces have always been of interest and revolutionary systems have recently been completed for Eton College and Trinity College in Cambridge. He also lectures regularly at the departments of architecture and engineering at the University of Cambridge. A keen musician, he is a member of The Bach Choir, the CUMS Chorus and Cambridge Voices and can still remember how to play jazz bass, a sure sign of a misspent youth. Recent hobbies include building Brio train tracks for honorary grandchildren and looking after strange percussion instruments for the University of Cambridge.

Abubakr Merghani, BSc, MPhil, PhD, SEC, ISA

Abubakr graduated from the University of Khartoum in 1993 with the top First Class Honours in Architectural Engineering. Gaining the Chevening Scholarship, he obtained his MPhil degree from the University of Cambridge in 1998 during which he was awarded, as part of a team, the First Prize in the Wates Built Homes International Design Competition towards sustainable housing for the twenty-first century. His PhD degree, Cambridge University 2001, discussed the thermal performance of traditional courtyard buildings in Sudan. In 2002, he took up his current position as Head of the Architecture and Design Department, ComputerMan College, Khartoum, Sudan. He also works privately on various design projects and is a member of the Sudanese Engineering Council (SEC) and the Institute of Sudanese Architects (ISA). His research interests include the environmental aspects of the built environment.

Marialena Nikolopoulou, BEng, MPhil, PhD

Marialena is an architectural engineer specialising in environmental design, with a PhD in 'Thermal Comfort in Outdoor Urban Spaces' from the University of Cambridge. She is currently a lecturer at the University of Bath, prior to which she was a research associate at the Centre for Renewable Energy Sources in Greece and a lecturer at the Hellenic Open University. She is a member of the Technical Chamber of Greece, as well as the Hellenic Association for Renewable Energy Sources. Her research interests concentrate on issues of urban microclimate, thermal comfort, bioclimatic design and rational use of energy in the built environment. She has participated in various EU-funded research projects, has worked as an environmental consultant on architectural projects and competitions, and has contributed to numerous articles in refereed journals and international scientific conferences. She was the coordinator of the large-scale, EU-funded project 'RUROS: Rediscovering the Urban Realm and Outdoor Spaces', as well as the editor of a technical publication under the Fifth Framework Programme, City of Tomorrow and Cultural Heritage, on bioclimatic design guidelines for urban spaces. Finally, she has been awarded the RIBA Trust Research Award and the Human Biometeorology Scientific Award of the International Society of Biometeorology.

Katerina Parpairi, DipArch, MPhil, PhD

Katerina is a practising architect trained in Greece (National Technical University of Athens, 1994) and specialising in environmental design (MPhil, 1995) with a PhD degree in light and visual comfort (1999) from the University of Cambridge. She has received awards and scholarships including the RIBA Trust Research Award. She lectures at the Hellenic Open University (postgraduate courses from 2000 onwards) and at the Department of Architecture in the University of Thessaly (2001–03). Since 1999, she has been a major partner of the architectural firm 'Ergastiri '73' where she works on restoration projects as well as the architectural-bioclimatic design of housing, office buildings, hotels, etc. Recent research publications include: 'The Luminance Differences index: a new indicator of user preferences in daylit spaces' (with Baker, Steemers and Compagnon in *Lighting Research and Technology*, 2002); and contributing author in *Daylight Design of Buildings* (Baker and Steemers, 2002).

André Potvin, MArch, PhD

André completed his PhD at the Martin Centre, University of Cambridge Department of Architecture on urban environmental diversity. He returned to Canada and, from 1997, his post-doctoral research has focused on northern urban microclimates. He is currently Professor at Laval University School of Architecture where he teaches and conducts research on environmental design. His most recent research grant from the Social Sciences and Humanities Research Council of Canada deals with the development of a global environmental comfort index integrating the human environmental adaptability theory. André founded GRAP (Groupe de Recherche en Ambiances Physiques) at Laval University dedicated to the integration of passive architectural environmental control systems in buildings. He has conducted several studies for projects such as the New Canadian Embassy in Berlin, the CDP (Caisse de Dépôt et de Placement) of Québec in Montréal and more recently the Wood Research Centre to be built on Laval University main campus. The CDP project received the 2003 Medal for Architectural Innovation from the Royal Architectural Institute of Canada.

Marylis C. Ramos, BSArch, MPhil

Marylis Ramos is an architect from the Philippines whose research interests include south-east Asian architecture, environmental design, urban planning and outdoor comfort. She completed her MPhil in Environmental Design in Architecture at the University of Cambridge in 2001 and is currently working on her PhD on 'Comfort in Urban Outdoor Spaces' at the University of Cambridge under the supervision of Dr Koen Steemers. She is also doing some collaborative work involving urban morphology and comfort for the RUROS (Rediscovering the Urban Realm and Open Spaces) project, which is being coordinated by Dr Marialena Nikolopoulou from CRES (Centre for Renewable Energy Sources) in Greece.

Elizabeth Shove, BA, DPhil

Elizabeth Shove is a Reader in Sociology at the University of Lancaster, having previously served as director of Lancaster's Centre for Science Studies and as deputy director of the Centre for the Study of Environmental Change. Elizabeth has extensive experience of research at the intersection of social science, energy and environment and is co-author (with Simon Guy) of *A Sociology of Energy, Buildings and the Environment* (2000) and more recently of *Comfort, Cleanliness and Convenience: the social organization of normality* (2003). Current projects include work on the future of comfort and on sustainable domestic technologies.

Maria S. Sinou, DiplArchEng, MPhil

Maria is currently reading for a PhD in Environmental Design at the Martin Centre of Urban and Architectural Studies at the Department of Architecture, University of Cambridge. Her research focuses on the relation of urban space forms and their thermal performance. From 1990 to 1994 she studied Interior Design in the Technical Educational Institution in Athens, and from 1995 to 2000 Architecture in the Aristotle University of Thessaloniki. During her studies, she worked in two architectural practices and was involved in several projects in Greece. Maria successfully completed the MPhil course in Environmental Design in Architecture at the Martin Centre in 2001. She has contributed in several international conferences, and currently teaches environment and construction at the Department of Architecture in Cambridge.

Mary Ann Steane, BA, Dipl, MPhil

Mary Ann is a lecturer in the Department of Architecture at the University of Cambridge. She lectures on environmental issues and is also a course director of the MPhil in Environmental Design. In the studio courses that she delivers the acknowledgement and integration of environmental design issues is always given particular emphasis. Her research marries technical analysis with a more historical perspective. At the PLEA 2000 conference she received a 'Best Paper' award for her presentation of 'Building in the climate of the new world: a cultural or environmental response?' The current focus of her research is on natural lighting strategies, and seeks to enable more precise communication about light quality, and thus to develop a more nuanced appraisal of natural lighting strategies in their broader topographic, climatic and programmatic context.

Koen Steemers, BSc, BArch, MPhil, PhD, RIBA

Koen is an architect and environmental consultant: he has practised architecture in Germany, The Netherlands and the UK and is a director of the consultancy Cambridge Architectural Research Limited. He is currently the director of the Martin Centre for Architectural and Urban Studies and a reader in the Department of Architecture at the University of Cambridge. His main research interest deals with environmental performance related to built and urban form with respect to energy, light, ventilation, comfort and perception. He is the current president of the international association PLEA (Passive and Low Energy Architecture). With over 100 publications in the field of environmental design, his recent books include: *Energy and Environment in Architecture* (with Nick Baker, 2000), *The Selective Environment* (with Dean Hawkes and Jane McDonald, 2002), and *Daylight Design of Buildings* (with Nick Baker, 2002).

Illustration credits

Alvar Aalto Museum, Photo collection; Martti Kapanen, photographer   13.3

Aerophoto   3.3

Centro Studi Giusippe Terragni, by kind permission   13.4

Crowther, P. (2000) 'Chapter 2: Building Deconstruction in Australia', in Task Group 39 *Overview of Deconstruction in Selected Countries*, CIB Report, Publication 252, August, CIB   5.6

Deximage   3.2

Fanger, P. O. (1970) Thermal comfort, New York: McGraw-Hill   4.2

FLC/ADAGP, Paris and DACS, London © 2004   11.1 and 13.1

Guedes, M. (2000) 'Thermal comfort and passive cooling in southern European offices', unpublished PhD thesis, University of Cambridge   4.3

Hocker, B.   3.1

Humphreys, M. A. (1978) *Outdoor Temperatures and Comfort Indoors*, Building Research Establishment, Current Paper 53/78, Watford: BRE   4.1

Kibert, C. and Chini, A. (2000) *Overview of Deconstruction in Selected Countries*, CIB Report, Task Group 39, August: CIB   5.4

Saywood, M.   3.4

Terragni, G. 'Relazione tecniche', *Quadrante*, 1936, 35(36): 55   10.14

Wernick, I. K., Herman, R., Govind, S. and Ausubel, J. H. (1996) 'Materialization and Dematerialization: Measures and Trends', *Daedalus, Journal of the American Academy of Arts and Sciences*, 125(3)   5.4

PART 1

Introduction

Chapter 1

Environmental diversity in architecture

Mary Ann Steane and Koen Steemers

Introduction

This book has a twofold aim: to establish a range of useful definitions of environmental diversity and to explore the role it plays in ordering and enriching architectural experience. In bringing together architectural research work that clearly identifies why environmental diversity is of significance and how it relates to design, it illuminates the potentially pivotal role played by environmental thinking at all stages of the design process. It is perhaps not surprising that many contemporary approaches to design development prioritise spatial design, but with the intention of prompting debate on how design ambitions ought to be framed, the discussion presented here questions the degree to which spatial and environmental design can ever be separated. In addition it underlines why environmental design guidance needs to change to reflect the idea that diversity is a fundamental design criterion alongside comfort.

Discussion of the environmental aspects of architecture and the way in which people interact with buildings is not as commonplace as some might assume — much recent architectural discourse has concentrated on aspects of construction or aesthetics, while the analysis and review of environmental strategy has received considerably less attention. Despite the obvious significance of programmatic issues to design it is surprising how frequently either occupation patterns or the views of occupants have been ignored, as if commentary on how buildings are inhabited somehow diverts attention from the finished artefact that is the building itself. Buildings, however, are hardly finished without their inhabitants and the activities that they pursue inside or around them. As in Rasmussen's seminal work *Experiencing Architecture* (1959), and Heschong's later *Thermal Delight in*

Architecture (1979), a detailed commentary on our perception of and engagement with the built environment is offered here. This book takes the position that the dynamics of the architectural environment is a key aspect of good design, yet poorly understood. An antidote to the misconceptions of optimum environmental performance or fixed criteria, it seeks to demonstrate why a richness of environmental variety is worth pursuing.

Though concerned with environmental science, this book deliberately avoids providing a purely technical view. A wide spectrum of approaches are offered that are mutually supportive rather than exclusive, with each section of the book demonstrating how an understanding of a particular context or environmental characteristic informs design in dynamic terms. Ranging from the individual's perception to the urban scale, and encompassing visual, thermal, aural and climatic characteristics, different aspects of the issue are covered as follows:

1   the way in which environmental objectives shape design;
2   the need to characterise different kinds of urban structure in terms of the level of environmental diversity that they provide;
3   the scientific evidence for how and why environmental diversity is exploited by building users;
4   how and why the idea of diversity ought to engender a new role for architects and engineers, new approaches to building design and indeed even new kinds of building;
5   the way in which the pursuit of diversity has informed or qualified design ambitions through extended discussion of a range of case-study buildings.

The aim is not to overturn current thinking completely. Rather the book attempts to review some of the problematic ambiguities and unintended consequences of that thinking, and to indicate how the pursuit of diversity can help to reframe design issues and prompt new ways of exploring, testing and communicating design strategies.

The first section of the book introduces the issues related to environmental diversity from four perspectives: a sociologist's, an architectural philosopher's, a scientist's, and a technologist's view. These chapters take broad and discursive views that provide a wider framework for the subsequent discussions. What follows is a combination of technical and theoretical chapters that are structured into sections related to the urban and intermediate scale, and the interior environment, before ending with a chapter related to design integration. The structure is thus as follows.

Introduction: 1. Environmental diversity in architecture
Framework: 2. Social, architectural and environmental convergence
                 3. The ambiguity of intentions
                 4. Human nature
                 5. Designing diverse lifetimes for evolving buildings

Commodity, firmness and delight

It was in the first century BC that Vitruvius first suggested that 'commodity, firmness and delight' stated the proper ambitions of architectural design. Debates have subsequently arisen over such issues as whether 'delight' is a matter of personal taste or whether 'commodity' can be reduced too easily to supplying the lowest measurable common denominator, but essentially the three areas of concern have been accepted ever since. From the Enlightenment onwards, however, these different aspects of design have become increasingly divorced from one another, with 'firmness' and 'commodity' usually approximated to structure and comfort respectively, and defined in technical terms; and 'delight' to aesthetics, as a matter of what can be drawn or conveyed visually. The tasks of defining and achieving structure and comfort have now frequently been handed over to technical experts, while aesthetics and spatial design have been made the responsibility of the architect. The desire to encourage sustainable development is, however, fostering a new interest in the relationship between 'commodity' and 'firmness', in as much as a new emphasis on the potential future adaptability of architecture involves asking how structural and environmental strategies may co-evolve with changes of use and reconfigurations of the building fabric. It is also worth underlining that, experientially speaking, 'delight' and 'commodity' are not possible to separate, particularly if the emphasis on visual matters implied by the definition of delight as aesthetics is ignored in favour of a broader acceptance of the range of stimuli that architectural experience provides. In this exploration of environmental diversity it makes sense to reassess the relationships between both 'delight' and 'commodity', and 'firmness' and 'commodity', and to examine the extent to which the three aspects are interlinked.

Commodity and delight: comfort and stimulation

We need to begin by asking how 'commodity' or comfort is defined and experienced. In thermal terms, comfort has been defined as 'that state of mind which expresses satisfaction with the thermal environment' (ASHRAE 1992). The definition goes on to say that comfort is the absence of thermal discomfort and a condition in which 80% of people do not express dissatisfaction. Interestingly, the International Standards Office (ISO 1994) defines comfort in purely technical terms and suggests that comfort is achieved when there is 'thermal neutrality', or in other words maintenance of the body's energy balance. Although both AHSRAE and ISO establish relatively narrow standards for thermal comfort, it is important to note the reference to 'state of mind' in ASHRAE's definition. This suggests that despite the narrow physiological definition, thermal comfort is, at least in part, a psychological phenomenon open to influence by variables other than thermal. Recent research has identified some of the reasons for the discrepancies between the laboratory-based comfort studies, which form the basis of both ASHRAE and ISO standards, and field-based research, which recognises the significance of behaviour, context and culture. It is thus all the more surprising that such standards are applied in all climates and for all building types.

A further deficiency in the technical definition of comfort is the reference to 'absence of discomfort', which assumes that the absence of stimulus is good. The absence of discomfort is a 'commodity' that might be prescribed, yet it omits the potential 'delight' that can be present with some degree of stimulus and contrast. Melville in *Moby Dick* evocatively describes this apparently paradoxical state of affairs, where the presence of discomfort engenders a stronger sense of comfort:

> We felt very nice and snug, the more so since it was so chilly out-of-doors; indeed out of bedclothes too, seeing there was no fire in the room. The more so, I say, because truly to enjoy bodily warmth, some small part of you must be cold, for there is no quality in the world that is not what it is merely by contrast. Nothing exists in itself. If you flatter yourself that you are all over comfortable, and have been so a long time, then you cannot be said to be comfortable anymore. But if, like Queequeg and me in the bed, the tip of your nose or the crown of your head is slightly chilled, why then, indeed, in the general consciousness you feel most delightfully and unmistakably warm.
>
> (Melville 1966: 66)

Thermal comfort conditions are usually sketched out by indicating the range of air temperature and relative humidity levels within which most people feel comfortable. Olgyay's bioclimatic chart is one such description, although it needs to specify that these conditions only apply if the subject is at rest, in office dress, out of the wind and in the shade (Olgyay 1963). Yet, neither the measurement of the thermal environment or the appreciation of thermal comfort is as simple as this in reality. The mean radiant temperature, for example, is a better measure of human comfort than air temperature, though more difficult to gauge accurately. In addition, as Kwok (2000) nicely summarises, acclimatisation to particular conditions (physiological adjustment) may lead to changes in the definition of comfort, while the possibility of altering the environment (behavioural adjustment) or an expectation of a lack of comfort (psychological adjustment) may lead to a greater tolerance for temperature variation.

In visual terms, the definition of comfort is complicated by the fact that our eyes adapt to the light that is available. Our comfort depends on the quantity of light available and the relative brightness of the different areas of the visual field that are in view. We struggle to sew in too little light, but too much light for long periods can be tiring. We suffer from glare if our eyes are asked to adapt to too great a range of brightness at any one time. In aural terms, comfort is even more difficult to define. Apart from the kind of noise that is so loud it is literally deafening, or so rhythmically insistent, like a warning signal, that it cannot be ignored, it is only possible to discuss aural comfort from a relative point of view. Silence in certain circumstances may be as disturbing as loud or distracting sound is in others, it all depends on what kinds of conditions are being sought.

How we evaluate and communicate ideas about comfort is even more complicated. We are in fact conscious of a comfortable environment only when we know we have found it, or are aware that we have abandoned it, or when we are specifically questioned on the subject. Usually we take it for granted — when we are comfortable we do not notice our environment at all. It is usually assumed without question that thermal comfort requires the thermal environment to be an even, stable temperature but, as Melville notes, to be actively aware that we are for the most part comfortable, paradoxically some part of us has to be uncomfortable. The fact that we do not notice stability, and we do notice change or difference, means that the separate analysis and discussion of comfort and stimulation is not straightforward.

Because we are constantly in league with our environment, if not always consciously so, in addition to seeking definitions of comfort, we need to take the issue of how we make ourselves comfortable more seriously, whether in thermal, visual or aural terms. Definitions of comfort in technical terms need to be illuminated by an examination of the issue from a different point of view: the relationship between behaviour and setting. A more dynamic perspective is required that considers how and why we orchestrate stability or instability environmentally and the degree to which stable rather than unstable conditions determine comfort (and/or stimulation). By reframing this issue it is to be hoped that a more nuanced articulation of the relationships between occupants and building ought to be possible, whether building-scale strategy, spatial design or local detail is in question.

Commodity and firmness: adaptability and permanence

Adaptability over time is as much a concern in the life of a room, as in the life of a building or the life of a city. In the short term and at room scale, the issue affects how users can interact with the building, whether by reconfiguring an element of the building, as say in opening a window, or by moving from one place to another within it, to create or find a more appropriate environment. In the medium-term, and at building scale, the issue that needs to be addressed is how the building's fabric can be adapted to respond to changes in conditions (e.g. working patterns, changes in the immediate physical or environmental context, or climate change). The pursuit of adaptability is directly linked to the provision of diversity in as much as buildings whose environmental conditions vary to embrace the potential of the local site and climate are more likely to accommodate a range of uses with only relatively small-scale transformations of the building fabric. But the question is more complex than this. It has been pointed out that more attention needs to be given to how building design may be made more indeterminate, that is to say, at the initial design stage, the range of possible future occupation patterns of a building needs to be taken more seriously. Clearer thinking on how 'commodity' and 'firmness' are in reciprocity with one another is required, whether this has to do with the range of lifetimes that ought to be projected for the different elements of construction, or the manner in which changes to the fabric might affect the environments that the building provides. This is a matter of understanding how to devise structural and environmental strategies that make buildings environmentally adaptable, so that they can be made more specific to particular uses. This contrasts with providing conditions that are excessively flexible and, though apparently neutral, are actually more difficult to fine-tune sensibly in spatial or environmental terms.

In the longer term and at the urban scale, the drive to make cities more sustainable demands a more in-depth study of urban dynamics, that is to say of the structures and processes that have informed urban order and allowed cities to accommodate change. In acknowledging the three general areas of urban order — physical structure, environment and civic ideals — such a study of the urban metabolism ought to ensure better communication between all those concerned with the social, environmental and economic issues that are involved, and allow them to collectively identify and nurture more sustainable patterns of development that balance the need for permanence with the need for evolution while maintaining the appropriately diverse range of environmental conditions without which the urban ecosystem cannot prosper.

The characterisation of architectural experience

As a diversity of environments is what buildings always provide, at least to some degree, how diversity can be a design criterion deserves exploration.

Although, to do this, what is meant by diversity itself clearly needs to be characterised, and thus the identification of a range of possible aims in designing for diversity is a valuable first step:

- to coordinate a dynamic environmental strategy with other spatial, programmatic or social intentions in order to ensure that the architecture provides a series of appropriate and stimulating settings and sequences that vary over time and/or space;
- the provision of adaptive opportunities that allow occupants to engage with the building and take active control of their environment;
- to orchestrate sequences or transitions, especially between the exterior and the interior (or vice versa), in appropriate ways at different times of the day or season;
- to moderate climatic conditions in such a way that occupants remain aware of the passage of time and weather, and can appreciate their distance from or intimacy with the vagaries of the climate;
- to take account of the range of possible future uses which a building may be asked to fulfil and to allow this to inform the permanence or impermanence (the projected lifetimes) of different elements of the building's fabric;
- to create a robust indeterminate urban framework within which a diverse range of spaces, environments and activities can occur and whose fabric is adaptable and thus can accept changes in occupation patterns over time;
- to allow occupants to articulate their own environmental preferences in accordance with local traditions, conventions and practices.

Environmental diversity is a design characteristic that is closely related to our experience of architecture. The conscious shaping of diversity, that is to say, the conscious orchestration of the dynamic patterns of environmental variation, is made possible by an appreciation of its spatial and temporal aspects. Variables such as heat, light or sound may fluctuate over time and/or space. These fluctuations may be more or less wild, more or less sudden, depending on such issues as the nature of the space, its occupancy pattern, the form of construction and the climate. Thermal, visual and aural environments may thus be more or less stable temporally, and more or less even spatially. Before considering how these different degrees of diversity might be established, a brief review of the basic spatial, material, contextual and climatic parameters that allow different degrees of diversity is necessary.

The thermal environment

> As with all our other senses, there seems to be a simple pleasure that comes with just using our thermal sense, letting it provide us with bits of information about the world around, using it to explore and learn, or just to notice.
>
> (Heschong 1979: 18)

The thermal environment that the body senses is determined by the combined environmental characteristics of air temperature, air movement, the radiant field and relative humidity. We can feel thermal asymmetries when they are present, such as the strong radiant heat from an open fire in an otherwise cold room, or the pleasant breeze on a warm day, yet as long as they are in balance, comfort can be achieved. Such thermal conditions and fluctuations are directly influenced by architectural space and materiality in relation to solar orientation, wind direction and response to diurnal temperature fluctuations. The basic form and configuration of openings will determine sunlight and air movement patterns, and the nature of the building materials will influence temperature patterns. Research has demonstrated that occupants sitting near windows express a greater degree of thermal comfort. This is in part because, assuming that windows and blinds are operable, they have more choice and can create a range of thermal conditions by opening the window to introduce cool air and increase air movement, or lowering blinds to provide shading from the sun. It is thus not a question of providing an optimum and constant thermal environment but of providing adaptive opportunities through design to create thermal diversity. The links between architecture and thermal conditions are evident. The dynamics of a thermal environment, and opportunities to adapt those conditions, demonstrate the notion of temporal diversity in relation to design.

The cathedral environment is a case in point, where a thermally heavy building envelope with small openings results in relatively constant internal thermal conditions. As temperatures outdoors fluctuate up and down, reaching a peak during the early afternoon, the experience of entering the relatively cool interior of the cathedral plays a key part in altering the state of mind of the visitor. They have left behind the fluctuations of the external conditions — and the hubbub of daily life — and enter a cool, stable environment for a more contemplative experience. This spatial contrast in thermal environments demonstrates again how diversity relates to design, which we will refer to as spatial diversity.

But even in the outdoor urban context there are diverse conditions: the choice of sun or shade, the evaporative coolth generated by a fountain or park, the windy street versus the protected arcade, etc. One can define broad general choices that people will make, such as at above about 25°C the majority of northern Europeans will tend to seek shade in their cities, whereas below that being in the sun is preferred. However, the critical issue is that a diverse and appropriate range of opportunities exists that allows people to adapt to the prevailing conditions and their activities. Biological and psychological research has shown that a diverse environment — one that presents the greatest degree of choice and widest range of conditions — is highly desirable.

The aural environment

> There is no longer any interest in producing rooms with differential acoustical effects — they all sound alike. Yet the ordinary human being still enjoys variety, including variety of sound.
>
> (Rasmussen 1959: 235–6)

The aural environment can be described in terms of a number of characteristics, extending beyond the semantic differentials of loud or quiet to include: reverberant or dead; favourable or annoying; constant or variable; background or foreground, etc. The combination of such characteristics is determined by architecture and establishes the soundscape of our environment that in turn affects our perception of the architecture. Thus, the hushed and reverberant environment of a church is appropriate for quiet contemplation and reverence, and contrasts with the noise of daily life in the piazza outside. This distinction is reinforced by thermal and luminous transitions to create an experience of architecture that goes beyond the spatial or formal. Similar contrasts can be created for concert halls, where the chatter of voices in a hard, voluminous foyer replaces the urban noise, before entering the acoustically absorbent interior of the auditorium.

Rasmussen's claim that modern interiors all sound alike is juxtaposed with his description of the Rococo town house, highlighting the role of contrasts and sequence in aural conditions.

> From the covered carriage entrance the visitor came into a marble hall which resounded with the rattle of his sidearms ... Now came a series of rooms with more intimate and musical tones — a large dining room acoustically adapted for table music, a salon with silk- or damask-panelled walls which absorbed sound and shortened reverberations, and wooden dadoes which gave the right resonance for chamber music. Next came a smaller room in which the fragile tones of a spinet might be enjoyed and, finally, madame's boudoir, like a satin-lined jewelry box, where intimate friends could converse together ...
>
> (Rasmussen 1959: 234)

Not only does this quote highlight the role of architecture in creating the aural environments appropriate to the functions, but it also suggests that the acoustic sequence of spaces, in this case from public to private, are important. One might extend the argument by suggesting that the anticipation of entering the marble hall or satin-lined boudoir will influence perception and behaviour. Furthermore, the sound of the carriage wheels on the gravel drive and clattering footsteps in the hall will resonate and be heard deep into the house. Thus the connectedness of the spaces and the perception of their diverse acoustic conditions allow us to speak of what Rasmussen referred to as 'hearing architecture'. Unlike thermal or visual perception, where a space can be sensed as cool or warm and dark or light, we hear the architecture only because we activate it by the noise we introduce — the footsteps or the spinet. And yet, noise is formally defined as unwanted sound.

The noise of the city, or noise in a bar, is part of our expectation and we would find it curious if not disconcerting not to hear it. Furthermore, noise in an office or restaurant provides the essential masking to create a sense of privacy, whereas very quiet conditions make every word of a conversation audible to others. The booth is on the edge of a voluminous, reverberant dining room that creates an acoustic environment of background noise, yet the proximity between diners and adjacent surfaces in the booth help to make a private conversation comfortable. The simple presence of a nearby, hard surface improves the speech intelligibility in the midst of a bustling restaurant. Clearly, too much background noise will erode speech intelligibility, so an appropriate balance needs to be struck. The level of background noise becomes a conscious environmental design decision related to the vibrancy that is intended. Because reverberation is a key parameter that determines noise level, and is a function of room volume and absorption characteristics of surfaces, it is clear that the choices of architectural characteristics related to space and materials are critical.

The noise that people create can be considered either as wanted or unwanted, depending on the context: in a cafe it is welcome whereas in a library it will not be. Thus the noise source, as well as its level, are significant variables. The modern obsession with prescriptive values and precision can lead to environments that are strictly isolated, offering no sense of anticipation or connectedness. An example is music practice rooms, where a slight relaxation in sound insulation level between room and corridor can offer the user of the building auditory glimpses into the use of the building: pausing to hear the music filtering through, making the building seem active. This kind of graduation of sound, or connectedness between spaces of acoustic diversity, makes choices perceptible, makes the building audible.

In generic terms, people express a preference for natural sounds (birds, water, etc.) over artificial or mechanical sounds (vehicles, construction, etc.). A fountain in a square can usefully mask unwelcome traffic noise by the more welcome sound of splashing water. If consciously designed such acoustic diversity can be particularly effective, for example where the sound of water is a transient experience between a noisy urban environment and the more peaceful conditions of a museum or an office foyer. The sound of splashing becomes an environmental threshold, as significant as, yet more subtle than an architectural threshold.

The visual environment

> Light is not perceptible without form — even the diaphanous form of swirling smoke — to reflect it. Conversely form is not perceptible without light to reveal it, at least not to our vision, on which we rely to provide the majority of our information about our surroundings.
>
> (Millet 1996: 47)

Five key factors determine the visual environment of a daylit space: 1) light quality; 2) spatial orientation and 3) geometry; 4) the size, configuration and glazing of openings; and 5) the specification of surface colour and texture. The physical context affects the quantity and character of light that reaches the building, but the diversity of the visual environment is also dependent on the climate, in as much as this dictates the dynamics of natural light available. As a constantly varying source, in both quality and quantity, natural light has an advantage over all artificial sources but candlelight, in that it changes with cloud cover and the time of day or season. Aside from cutting energy costs, research has established that the use of daylight is generally preferred for this reason. Whether the available light is predominantly diffuse skylight or direct shadow-casting sunlight makes a significant difference to the character of a naturally lit interior, with more diffuse light tending to produce more softly lit, less harshly modelled, less spatially dynamic visual environments and more direct light having the reverse effect.

At a basic level it is, of course, the diversity in levels of brightness across the visual field that allows us to see our surroundings, but the careful handling of visual diversity by designers is important for a number of reasons. At a local level, research has confirmed that in many situations the creation of visual interest helps to establish appropriate conditions for many different activities. Contrasts in brightness between background and task can help to focus attention on a particular visual task or help to create particular territories within a space, while distinctive contrasts in colour can allow particular objects or surfaces to recede or become more visually prominent. At the scale of the building, visual diversity may be consciously deployed to create thresholds or transitions in light, to engender spatial drama or to guide movement.

Indices of diversity: orchestrating and ordering dynamics

The embrace of diversity as an environmental design criterion ought to prompt new interest in how indices of diversity can be established. The discussion above sets out the different physical parameters that help to determine the characterisation of thermal, aural and visual diversity. While each deserves individual attention, it is hoped that an attempt to outline visual diversity will throw light on how the broad pursuit of environmental diversity ought to inform design.

The spectrum of possible levels of visual diversity ranges from those environments that are most stable and most even and that offer little visual contrast, to those that are least stable, least even and visually complex. In experiential terms, visual stability/evenness might be characterised as calm, austere or even dull, and visual instability/differentiation as rich, animated or visually noisy. Intermediate conditions might be more dispassionately described as safe or neutral, or more evocatively described with adjectives like measured, fresh or aqueous.

Any of the spatial/environmental moods outlined above might be appropriate (or inappropriate) depending on the particular occupation pattern of a space. Designing with diversity in mind should not mean that spaces with low diversity are never possible or that spaces with high diversity are introduced without question. While it is obvious that some activities are definitely enhanced by calmness as opposed to distraction, and vice versa, how any individual space contributes to the character of environmental sequences across space and time needs to be carefully considered. Either a calm or a complex space may be able to provide interest to a sequence of spaces through contrast. To put this another way, at the building scale the provision of a greater level of diversity and the provision of an appropriately stimulating environment are not always the same thing. Individual spaces may have low diversity but nevertheless contribute to the spatial diversity of the whole in a way that can be experienced by those moving through the building. It is the orchestration of diversity at the building or landscape scale that needs to take into account all the experiential implications of any environmental strategy: whether to do with stimulation of the senses, the provision of a range of appropriate (or appropriately adaptable, or appropriately robust) conditions; or the qualification and fine-tuning of relationships between building users and their physical and climatic context.

In any climate 'normal' conditions are usually provided by an intermediate level of diversity. In establishing the kind of environment that would not usually be noticed, these are the conditions with which we are familiar and thus comfortable, and which tend to allow architecture to be read as neutral, as the background. The idea that safe, comfortable environmental strategies lie somewhere in the middle range of an index of diversity and that potentially more memorable, more striking environments lie at either end should not be surprising. In prompting greater focus on how potentially complex patterns of environmental variation can be structured, indices of diversity ought to help designers grasp the degree to which any particular strategy needs to differ from normal conditions more firmly, that is to say, how and why it should make more of a claim on attention. They have the additional benefit of not allowing comfort and stimulation to be seen as polar opposites, instead helping to underline the extent to which the conditions of comfort can provide stimulation and the reverse, whether understood from the perspective of spatial or building-scale environmental strategy.

Environmental diversity can be discussed in at least two ways: spatial diversity and temporal diversity. Spatial diversity relates to the presence of a range of environmental conditions that can be formally structured as part of a sequence. The transitions between such spaces — whether gradual or sharp — will influence our perception of diversity and the opportunity to choose or anticipate contrasting conditions. This connectedness of contrasting environments is therefore a key characteristic that determines the perceived spatial diversity. One might thus propose that increased connectedness and contrasts result in greater diversity, whereas a lack in either connectedness or contrast results in low diversity.

Temporal diversity relates to a specific place, interior or urban, and how it changes or is changed over time. Thus a space can be stable or unstable in environmental terms, where instability results in greater temporal environmental diversity. Temporal diversity will clearly have an impact on spatial diversity, as contrasts between spaces will change over time. Thermally a space is stable when thermally massive like a cave as compared with a lightweight glasshouse. Acoustically a space is diverse if it is reverberant and open, and stable when absorptive and sealed. A space is visually diverse if finishes are matt and no direct light is available, as opposed to a room with glossy surfaces and where sunlight is allowed to enter. Similarly, a space that offers significant adaptive opportunities to the user can be defined as environmentally diverse. At a finer scale, spatial diversity can occur within a larger space: thermal, acoustic and luminous variations may be different within the same space. An example of this is how progress through a space can change our visual perception of it: this is discussed in more detail later in the book.

Architecture is frozen music?

> Architecture is not produced simply by adding plans and sections to elevations, it is something else and something more. It is impossible to explain precisely what it is — its limits are by no means well-defined. On the whole, art should not be explained; it must be experienced.
>
> (Rasmussen 1959: 9)

Although it is generally agreed that as an art, architecture demands a bold imagination, a good technical understanding, and good judgement as to how design conflicts and intentions can be addressed and resolved, to explain what drives architectural thinking is difficult without turning to the other arts for parallels. Such parallels can be very illuminating yet they may also obscure key insights into the nature of the discipline. The phrase 'Architecture is frozen music' is a case in point. Like music, architecture orders experience. Yet while sequence, repetition, tempo or phrasing, all have architectural equivalents that support the basic analogy, the emphasis on the idea that, in contrast to music, architecture is stable or silent, is mistaken. This kind of analogy assumes that architecture is fundamentally an artefact, a construction. While this would usually go unquestioned, it is a definition that helps to obscure one of architecture's other basic concerns. It is equally valid after all to suggest that architecture is the shaping of shelter, the provision of particular environments. When considered in these terms, architecture is never frozen. It is fundamentally dynamic, changing more or less slowly, more or less quickly, with alterations in use or the climatic and physical context. It is this dynamic quality of the built environment that stimulates our senses, yet it is rarely anticipated or understood in the design process due to an emphasis on the geometric and physical aspects of design that are represented by the kind of architectural drawings to which Rasmussen refers. As Rasmussen also points out, it is not easy to explain either what architecture is or how architectural decisions are arrived at without taking architectural experience seriously. It is the order of this dynamic behaviour, the measurement of experience, that this research seeks to illuminate.

References

ASHRAE (1992) *Thermal Environmental Conditions for Human Occupancy*, ASHRAE-55, American Society of Heating, Refrigeration and Air-conditioning Engineers.

Heschong, L. (1979) *Thermal Delight in Architecture*, Cambridge, Mass.: MIT Press.

ISO (1994) *Moderate Thermal Environments: Determination of the PMV and PPD indices and specification of the conditions for thermal comfort*, ISO 7730, Geneva: International Standards Organization.

Kwok, A. G. (2000) 'Thermal boredom', in Steemers, K. and Yannas, S. (eds) *Architecture + City + Environment, Proceedings of PLEA 2000*, London: James & James.

Melville, H. (1966) *Moby Dick*, London: Reader's Digest Association.

Millet, M. (1996) *Light Revealing Architecture*, New York: Van Nostrand Reinhold.

Olgyay, V. (1963) *Design with Climate: Bioclimatic approach to architectural regionalism*, Princeton: Princeton University Press.

Rasmussen, S. E. (1959) *Experiencing Architecture*, Cambridge, Mass.: MIT Press.

PART 2

Framework

Chapter 2

Social, architectural and environmental convergence

Elizabeth Shove

Introduction

Tensions between variety and standardisation ripple through architectural discourse and are capable of arousing strong emotion. From some perspectives, even limited initiatives in modularisation, prefabrication and technical regulation threaten designers' creativity, innovation and invention and so limit the richness of the built environment. More ordinarily, the spectre of bland uniformity is, time and again, set against the thrill of difference, the drama of sensation and the delight of contrast. On the other hand, there is an equally compelling tradition of optimisation, of organising and controlling variation with the entirely reasonable aim of producing structures that satisfy as many of their users as possible. Indeed, this ambition is central to the very project of architecture.

There are, of course, multiple dimensions of variety and standardisation and many points of reference, perspective and value. Rather than unpicking these issues in the abstract, I want to explore elements of this ambivalence with reference to the social and architectural construction of the indoor environment. This is a good case to take on a number of counts. In simple terms, people come in many shapes and sizes and, as decades of thermal comfort research has shown, their physiological 'needs' vary widely. Since most buildings are designed to cater for more than one person, the challenge of managing such variation is unavoidable. Second, a number of commentators have argued that peoples' experience of what Lisa Heschong (1979) refers to as 'thermal delight' — that is their positive enjoyment of the indoor and outdoor environment — depends upon the practical experience of contrast: of moving between one set of conditions and another.

In this chapter I argue that efforts to respond to the first of these dimensions, that is to cope with physiological differences between people, have had the seemingly paradoxical effect of eroding differences at the level of individual and cultural experience. The net result appears to be one of social, architectural and environmental convergence and, in Heschong's terms, a diminution of thermal delight in the name of meeting peoples' needs. This is perhaps to be expected. After all, people are full of contradictions, architecture is full of compromises and in this case there really are competing approaches.

But by representing the problem as one of determining and meeting users' needs — whether those be for optimally standardised conditions, for personal control, or for contrasting sensations — architectural and scientific commentators miss the rather important point that all of these 'needs' have a social and cultural history. More than that, such histories are profoundly marked by the built environment itself. Adding these threads together, I conclude that in responding to the challenge of meeting peoples' needs (whether through optimisation of the indoor environment or through the provision of personalised controls and 'adaptive opportunities') designers are at the same time involved in making expectations and in 'scripting' what are becoming increasingly standardised social practices.

This dynamic is of some significance for the global environment. For reasons detailed below, the past century has witnessed an impressive but largely unnoticed convergence of indoor environmental conditions around the world. Local and seasonal differences are being ironed out through widespread reliance on standardised materials, technologies, design guides and building codes. Peoples' experiences of indoor conditions and hence their interpretations of comfort are converging as reliance on unsustainably energy-intensive forms of mechanical climate control becomes the norm. This raises important questions. Can architecture help to turn back the clock of global environmental change? Is it possible, through design, to deliberately re-engineer more diverse (and so less resource-intensive) cultural expectations of comfort? Might designers wilfully contravene current social and engineering convention, producing buildings that deliberately challenge and extend concepts and experiences of comfort in a conscious effort to overthrow contemporary regimes of 'thermal monotony' in the cause of sustainability?

This is controversial territory but it is territory around which discussion of standardisation and diversity inevitably revolves. In exploring such issues in more detail, the following sections make the case for thinking again about architects' and designers' political and economic roles in actively and creatively constructing the needs — be they for variety or for optimisation — that they seek to meet.

Valuing difference

I begin by highlighting three relevant but contrasting interpretations of the importance of difference regarding peoples' relationship to the thermal environment.

Human variation constitutes the basic starting point for a dominant family of ideas about thermal comfort. The so-called heat balance model describes the physical relationship between a person and his or her environment. Defined thus, comfortable 'neutrality' is that state in which the heat generated by the human body is equal to the heat transferred away. As Brager and de Dear explain:

> Heat balance models view the person as a passive recipient of thermal stimuli and are premised on the assumption that the effects of a given thermal environment are mediated exclusively by the physics of heat and mass exchanges between body and environment.
>
> (Brager and de Dear 1998: 84)

Since the physical properties of peoples' bodies vary significantly with size, shape, etc. so does their relation with the surrounding environment. Conditions ideal for the thermal physiological 'needs' of one person will, by definition, not suit those of someone who is smaller or larger or who has a different metabolic rate. In this analysis, the meaning of comfort is constant (here defined as thermal neutrality): what varies and so what has to be controlled in order to deliver comfort, is the relation between the body and its environment.

The second family of ideas takes comfort to be a continual achievement, a question of sensation and meaning and, as such, not a uniformly specifiable state of affairs. If there is any one constant, it is the need for variety, that being the spice of life. Picking up this theme and developing it in an unashamedly romantic manner, Heschong (1979) equates thermal variation in architecture with delight. She argues that thermal pleasure is a quality of changing experience. The tingling sensations of coming into the warm on a cold winter's day or the shock of stripping off and plunging into a cool pool in the height of the summer exemplify such transitions. Less dramatically, heating technologies like the traditional stove afford their users a variety of thermal possibilities: people can gather close to get warm and back off again when they get too hot. Such to-ing and fro-ing is simply not an option for those who inhabit mechanically controlled environments designed to deliver uniform conditions throughout. In suggesting that standard conditions of this kind constitute a form of sensory deprivation, Heschong subscribes to a distinctive theory of comfort in which difference (within limits) is of the essence. Likening thermal to culinary experience, Heschong draws parallels with the construction of a gourmet meal in which the taste and texture of each contrasting or complementary course enhances the sensation of the next. She bases her account of thermal delight on an assortment of cultural evidence — on what people say about what they enjoy — and on a more ideological opposition to standardisation. She writes as follows:

> the thermal environment … has the potential for such sensuality, cultural roles and symbolism that need not, indeed should not, be designed out of existence in the name of a thermally neutral world.
> (Heschong 1979: 16)

Heschong's analysis focuses on individual experience of difference, not on individual differences in terms of physiology. However, individual experience is not taken to be random or idiosyncratic for, in developing her argument, Heschong acknowledges patterns of cultural diversity. As she explains, the process of achieving comfort is a social one and one that societies have traditionally managed in significantly different ways. She refers, for instance, to the Greek tradition of an evening promenade, a collective event in which families and friends stroll together enjoying the relative cool after the heat of the day. To give another example, the Japanese convention of the 'kotatsu' — a heating device around which family members gather to keep warm — has all sorts of implications for domestic sociability (Wilhite et al. 1996).

Compared with these arrangements, the development and widespread use of mechanical heating and cooling systems tends to disrupt existing traditions and ways of life and in the process engender new, more homogenous concepts of comfort. For commentators like Heschong, trends of this kind represent a loss of meaning and symbolic togetherness as well as a diminution of personal physiological satisfaction. Whether one agrees with this judgement or not, the basic proposition is that understandings of comfort are the outcome of the culturally and historically specific means of its achievement. As a result, its specification is both contingent and inherently malleable.

In practice, these three aspects of difference — the physiological, the experiential and the social — have distinctive practical consequences for the design and management of the indoor environment.

Coping with difference

The fact that people, seen as thermal-biological systems, vary so widely has represented something of a problem for architects, engineers and manufacturers of heating and cooling systems. Having developed the capacity to reproduce just about any indoor weather system, designers 'became curious about what a truly optimal thermal environment might be' (Heschong 1979: 15). This was no idle curiosity for the challenge was — and still is — to produce solutions which do the best possible job of meeting most peoples' needs, most of the time. Subsequent efforts to define what these needs really are embody a range of methodological, philosophical and commercial preoccupations, some of which are outlined below.

In writing about the early history of air conditioning, Gail Cooper (1998) highlights the range of 'ready-made' natural models on offer. Should engineers aim to replicate idealised environments like those of a fine spring day in the mountains or was the more perfect climate that of a summer afternoon by the seaside? If the latter, the technological challenge was clearly one of 'bringing the best of the beach indoors' (Cooper 1998: 77). Such options were the subject of heated debate during the 1920s and 1930s, especially since the very idea of creating an artificial environment ran counter to the well-established philosophy of 'fresh air'. As Cooper explains, this kind of definitional instability was decidedly unhelpful in constructing markets for the fledgling air-conditioning industry.

Fortunately, nature also provided a less ambiguous point of reference. By turning from the analysis of a multiplicity of natural environments to the detailed study of human physiology, scientists employed in the research laboratory of the American Society of Heating and Ventilation Engineers sought to specify key parameters like those of temperature, humidity and air movement, and to thereby define human comfort precisely and in quantitative terms.

It is perhaps no accident that the science of physiology became the science of comfort or that highly controlled laboratory studies continue to provide the evidence on which current understandings depend. The process of experimentally isolating and reproducing a range of environmental parameters and recording subjects' responses to these conditions is a form of enquiry ideally suited to the analysis of variety and, through that, the specification of optimal conditions (Fanger 1970). Just as systematic recording of human variation allowed medical science to determine the properties of the statistically 'normal' child (Armstrong 1983), so thermal comfort researchers have plotted and predicted average responses to different conditions. In addition, this process of statistical optimisation depends upon the capacity to modify relevant variables at will. This is relevant, for the ability to implement and act on the resulting conclusions presumes a corresponding measure of mechanical control. Last, but by no means least, the methodology of averaging and normalisation itself ensures that few naturally occurring and/or naturally variable conditions will match up to the ideal thus defined. Gail Cooper spells out the commercial implications as follows:

> When natural climate was the ideal, mechanical systems sometimes fell short; but when quantitative standards of human comfort became the measure, natural climate was found wanting.
> (Cooper 1998: 79)

More than that,

> when it was shown that no natural climate could consistently deliver perfect conditions, air conditioning broke free of its geographic limits. When no town could deliver an ideal climate, all towns became potential markets for air conditioning.
>
> (Cooper 1998: 79)

There are many ways in which 'needs' and meanings of comfort might be determined (Humphreys 1995). As the above paragraphs indicate, the dominant method, now enshrined in codes and standards that are used around the world (Janda and Busch 1994), relies on a form of scientific analysis that has led to the precise specification of 'ideal' conditions through detailed investigation of what are taken to be the universal properties of human physiology. Cultural differences are simply overruled by this model, a feature that is of positive value in constructing a truly global market for a uniquely standardised concept of what people want.

Meeting and making needs

This far, I have gone along with the idea that peoples' needs exist and that what differs is how they are determined, analysed and met.

As described above, significant scientific and commercial resources have been invested in determining and reproducing conditions of comfort, routinely defined as a steady physiological state. Some argue with the details of this specification, for instance marshalling evidence to show that people 'need' thermal stimulation and variation and claiming that the standardising methodologies of laboratory science have failed to capture important dimensions of human experience. Such arguments suggest the need for further work to determine the nature and degree of environmental variation that human beings 'really' require, or to detail the extent of their psychological and biological adaptability (Brager and de Dear 2000; Nicol and Humphreys 2001).

The implication here is that the physiological and psychological quest for comfort should continue and should take account of more variables, incorporating or at least testing the idea that people have a biological need to experience different conditions and exploring the possibility that adaptation is normal and is, in fact, part of being comfortable. Such a route leads toward a science of variation and a different style of optimisation compared to that with which we have become accustomed (I will have more to say about the practical and commercial implications of such a strategy later). Nonetheless, the bottom line for what we might term 'adaptive' science remains that of determining where the social and psychological boundaries of comfort lie so as to ensure the ever-more efficient and effective delivery of such conditions. Roberts' (1997) history of the Chartered Institute of Building Services Engineers, appropriately entitled *The Quest for Comfort* is a good illustration of this line of thinking.

There are, however, other more relativistic ways of conceptualising the issue. John Crowley's (2001) historical study, appropriately entitled *The Invention of Comfort*, shows how the concept of comfort — initially understood as a state of spiritual well-being — became 'materialised' during the seventeenth century. Once defined as an attribute (for example, of chairs, clothes, food or indoor environments), it became possible to debate degrees and qualities of comfort in a manner that was previously inconceivable. Since the seventeenth century, and across different cultures, specific interpretations of what those qualities are have proved to be impressively malleable.

Those who have studied thermal comfort in the field as opposed to the laboratory have, for instance, found that people report being comfortable under remarkably different conditions and under conditions that fall way beyond the margins of physiologically derived comfort zones. To give just a few examples, the Pakistani workers included in Fergus Nicol's study claimed to be comfortable at temperatures of up to 31°C. (Nicol *et al*. 1999: 271). At the other extreme, people have reported being comfortable indoors at around 6°C during an Antarctic winter (Goldsmith 1960). Focusing on European differences, more recent research described by Stoops indicates that Portuguese office employees are content with a much wider range of seasonal variation (up to 5°C) than Swedes who do not expect indoor environments to waver by more than half a degree (Stoops 2001). As Humphreys' (1976) catalogue suggests, this list could go on and on.

This sort of evidence begs what is at heart a sociological question, namely, how are we to understand the reinvention of comfort defined as a collective socio-cultural construct. Following Crowley's approach, the challenge is one of figuring out how and why different cultures and societies subscribe to specific concepts of comfort and how these ideas and practices change over time. This is an especially relevant task if, as seems to be the case, conventions are converging around the world and are doing so in ways that are ultimately unsustainable. As already mentioned, one intriguing possibility is that rather than 'meeting' pre-existing needs, built environments are actively implicated in creating them. I take this idea further in the next section on constructing convergence.

Constructing convergence

In a much cited article, 'The de-scription of technical objects', Madelaine Akrich (1992), considers the possibility that material objects frequently embody scripts or narratives of use. For example, the circular openings in a bottle bank are ideal for admitting empty bottles and jars but awkward if one wants to dispose of a sheet of glass (de Laat 1996). The bottle bank silently but effectively 'speaks' to its user: it likes bottles, it doesn't like panes of glass. More than that, it organises and orchestrates its users' actions: it 'tells them' what to do. Though resistance is still possible, users are more or less obliged to follow the script this object provides. Such observations clear the way for a more philosophical rethinking of the relation between humans and what Latour refers to as 'non-human actors' (Latour 1992).

Developing these ideas in a way that is extremely germane to the analysis of architecture, Latour laments the tendency for the social sciences to overlook the contributory and constitutive role of the 'missing masses', that is the hardware not just of bottle banks but also of doors, windows and forms of urban design in shaping and reshaping social, or as he terms it, sociotechnical, practice. In essence, his argument is that conventions and practices are structured by 'non-human' as well as by human actors.

Do buildings 'script' their inhabitants' understandings of comfort? In describing the intersection of social and technical expectation, Michael Humphreys subscribes to just such a view. He writes as follows:

> If a building is set, regularly, at, say, 22°C the occupants will choose their clothing so that they are comfortable at that temperature. If enough buildings are controlled at this temperature, it becomes a norm for that society at that period of its history, and anything different is regarded as "uncomfortable", even though another generation might have preferred to wear more clothing and have cooler rooms, or to wear less clothing and have warmer rooms.
>
> (Humphreys 1995: 10)

This introduces a further question and one that has already been at least partly addressed. If buildings script and thereby construct needs, what determines the specification of requirement? Of all possible interpretations of comfort how is it that one definition becomes the norm? And why is it that contemporary definitions foster the reproduction of uniform conditions around the world?

As observed above, the standardising science of comfort has proved crucial in creating a global market for air-conditioning and for other forms of mechanical heating and cooling. One might even conclude that, whatever the stated claims of meeting needs, buildings and comfort standards have been designed to provide suitable 'homes' for technologies as much as for people. This far, such interests have favoured the cultural and material reproduction of standardised definitions of comfort and the construction of correspondingly standardised markets.

However, that need not remain the case. Increasingly sophisticated technologies of control, and the capacity to tailor conditions and even 'personalise' air (Fanger 2001), appear to be creating the space — and the need — for new styles of physiological enquiry. Specification of what Baker and Standeven (1992) refer to as 'adaptive opportunity' will, for instance, require new psycho-biological understandings of the micro-dynamics of comfort. At a stretch, one might even imagine a future in which environmental pressures are such that designers and research scientists are encouraged to study and emulate past cultures of comfort, perhaps reinventing the siesta or rediscovering culturally specific sociotechnical systems for managing the indoor environment. The point here is that the future of comfort is open: meanings are malleable and concepts, technologies and practices may unfold in any number of different directions. Whether these routes veer in the direction of standardisation and homogeneity or whether they encompass (or reinvent) cultural diversity depends, I suggest, not on the more precise understanding of human biology but on the political economy of construction.

Conclusions

In drawing this chapter to a close I want to highlight three features. I have argued, first, that the very reasonable goal of meeting peoples' needs has led, in the case of thermal comfort, to the specification of optimal solutions based upon systematic analysis of individual (physiological) variation. When embodied in the fabric of the built environment, these standardising scientific conclusions have acquired a new form of social significance. In determining what is provided as normal, and thus what people become accustomed to, they have reformulated social expectations and led, in the longer run, to a commonality of global experience and a measure of cultural convergence. This is not an especially unusual example. In this case, as in many others, optimising strategies for coping with variety have the, perhaps unintended, consequence of eroding social and cultural diversity. Designers have been party to this process, relying on ergonomics and other branches of science to determine exactly what it is that people 'need', whilst also recognising that necessarily universalising conclusions often run counter to local traditions, conventions and practices.

Second, I have noted that there are commercial as well as scientific pressures to pursue the route of averaging and optimising. Although money can be made of variety — indeed consumer culture is replete with examples of increasingly precise distinctions and details of differentiation — the first step in commodifying the indoor climate has been to establish deficiency and need by specifying an ideal that can only be achieved through reliance on mechanical systems of heating and cooling. The next step may be one of developing a more tailored or customised approach to indoor environmental management. This might well be the perfect design solution. By offering occupants the chance to determine their own microclimates, designers place responsibility for the specification of comfort as well as for sustainability and energy consumption firmly in the hands of the user. However, this line of reasoning bypasses the more important question of what kinds of environments people of different cultures might aspire to and what conditions they might seek to reproduce when given the 'choice'.

Again we confront the basic question of where peoples' senses of thermal and indoor environmental 'normality' come from. In thinking this through it is clear that the built environment does more than meet or respond to pre-existing needs. It also offers some but not other possibilities and in that way structures the menu of options from which 'choice' is possible. In addition, it helps to construct conventions, playing an active part in formulating senses of normality and so structuring peoples' responses when confronted with what might seem like unrestricted opportunity. This begs the further question of how architectural conventions are themselves shaped and formed. In this chapter I have highlighted the importance of particular forms of scientific enquiry and drawn attention to the political and economic significance of such work in framing concepts of comfort. Consistent with this line of argument, it is unlikely that the political economy of comfort will remain the same. By implication, theories and understandings of what people 'need' (whether that be variety or uniformity) will co-evolve with the means and technologies of provision.

References

Akrich, M. (1992) 'The de-scription of technical objects', in Bijker, W. and Law, J. (eds) *Shaping Technology/Building Society*, Cambridge Mass: MIT Press.

Armstrong, D. (1983) *Political Anatomy of the Body: Medical Knowledge in Britain in the 20th Century*, Cambridge: Cambridge University Press.

Baker, N. and Standeven, M. (1995) 'A behavioural approach to thermal comfort assessment in naturally ventilated buildings', Proceedings, *CIBSE National Conference*, Eastbourne: 76–84.

Brager, G. and de Dear, R. (1998) 'Thermal adaptation in the built environment: a literature review', *Energy and Buildings*, 27: 83–96.

Brager, G. and de Dear, R. (2000) 'A Standard for Natural Ventilation', *ASHRAE Journal*, October: 21–27.

Cooper, G. (1998) *Air Conditioning America: Engineers and the Controlled Environment, 1900–1960*, Baltimore: Johns Hopkins University Press.

Crowley, J.E. (2001) *The Invention of Comfort*, Baltimore: John Hopkins University Press.

de Dear, R. (1994) 'Outdoor climatic influences on indoor thermal comfort requirements', in Oseland, N. and Humphreys M. (eds) *Thermal Comfort: past, present and future*, Watford: Building Research Establishment.

de Laat, B. (1996) *Scripts for the Future*, University of Amsterdam.

Fanger, O. (1970) *Thermal Comfort — Analysis and Applications in Environmental Engineering*, Copenhagen: Danish Technical Press.

Fanger, O. (2001) 'Human requirements in future air-conditioned environments', *International Journal of Refrigeration*, 24(2): 148–53.

Goldsmith, R. (1960) 'Use of clothing records to demonstrate acclimatisation to cold in man', *Journal of Applied Physiology* 15(5): 776–80.

Heschong, L. (1979) *Thermal Delight in Architecture*, Cambridge, Mass.: MIT Press.

Humphreys, M. (1976) 'Field studies of thermal comfort compared and applied', *Building Services Engineering*, 44: 5–27.

Humphreys, M. (1995) 'Thermal comfort temperatures and the habits of Hobbits', in Nicol, F., Humphreys, M., Sykes, O. and Roaf, S. (eds) *Standards for Thermal Comfort*, London: E & F N Spon.

Janda, K. and Busch, J. (1994) 'Worldwide Status of Energy Standards for Building', *Energy*, 19(1): 27–44.

Latour, B. (1992) 'Where are the Missing Masses? The Sociology of a Few Mundane Artifacts', in Bijker, W. and Law, J. (eds) *Shaping Technology/Building Society*, Cambridge Mass.: MIT Press.

Nicol, F., Raja, I., Allaudin, A. and Jamy, G. (1999) 'Climatic variations in comfortable temperatures: the Pakistan projects', *Energy and Buildings*, 30: 261–79.

Nicol, F. and Humphreys, M. (2001) 'Adaptive thermal comfort and sustainable thermal standards for buildings', in *Moving Thermal Comfort Standards into the 21st Century: Conference Proceedings*, Windsor, UK, 5–8 April 2001, Oxford: Oxford Centre for Sustainable Development.

Roberts, B. (1997) *The Quest for Comfort*, London: Chartered Institute of Building Services Engineers.

Stoops, J. (2001) *The Physical Environment and Occupant Thermal Perceptions in Office Buildings*, Department of Building Services Engineering, Gothenburg: Chalmers University of Technology.

Wilhite, H., Nakagami, H., Masuda, T., Yamaga, Y. and Haneda, H. (1996) 'A cross-cultural analysis of household energy-use behaviour in Japan and Norway', *Energy Policy*, 24(9): 795–803.

Chapter 3

The ambiguity of intentions

Peter Carl

Introduction

The concept of environmental diversity is intended not only to make environmental design responsive to a richer, and therefore more realistic, spectrum of criteria; it also seeks to contribute to a more profound cultural issue. Research into sustainability has opened a view of urban and architectural processes that are dynamic in nature, of considerable complexity of scale and inter-relatedness, and often with very long-term consequences. Moreover, the moral implications of sustainability have made of the environmental sciences a discipline capable of illuminating the nature of the whole, without, however, considering themselves to be moral disciplines. This chapter seeks to clarify the resulting ambiguity — the difference between moral or political intentions and technical objectives — by, first, establishing the limits of the science–humanities contest in which this issue is usually framed, and then by concentrating upon what can be learned from practice. Because it remains the point of departure for all subsequent interpretation, Aristotle's explication of the relation of the practical life to any higher understanding is taken as a basis for grasping what is at issue. The main purpose of this chapter is to attempt to discover what is required to turn the ambiguity into an opportunity.

The good city: culture and energy

Two things which pertain to the whole of a city are culture and energy use. One may think of a city as the topographic distribution of culture and energy use. Moreover, it would seem obvious that energy is expended for the sake of culture, although we will see shortly that this is obscure in current practice. The concern for sustainability has made energy use part of moral concern. This creates an interesting and potentially fruitful ambiguity: on the one hand, the merits of sustainability are evident, on the other, no one would strive to subject culture to the sort of optimisation criteria normally deployed in, for example, environmental or engineering analysis. The ambiguity turns on the meaning of good in moral or political discourse and in that of technical discourse. When one seeks to realise moral or political goods through technical means, it may be termed an 'ambiguity of intentions'. One might refine this with regard to the difference between intentions and objectives. It may be a cultural intention, for example, to build a library because it is believed that knowledge, wisdom and creative citizenship go together. For an environmental scientist, this intention must be translated into an objective, such as comfort, which can be specified in the terms of the physical sciences.

The terms 'culture' and 'energy', in their current uses, are both products of the Enlightenment, and communication between them is marked by the divide between the humanities and the sciences. Whilst there is a long and still-evolving effort to model the humanities on the natural sciences — as expressed in such phrases as 'the human sciences' — the reverse is much more rare. This effort is generally conducted at the epistemological level, according to protocols having their source in Descartes. The wealth of knowledge produced according to these protocols is reckoned to be among the achievements of culture — which comprises centuries of cultural evolution (that is, design and production of a computer chip does not happen *ab nihilo*, but requires the Industrial Revolution, as well as all its research, equipment, projects, wars, etc.). At the same time, this particular (technological-capitalist) cultural history has been the object of regular, if ineffective, protest from the humanities; and it is precisely the cultural critique contained within sustainability that has reawakened speculation of a more coherent understanding and more fruitful collaboration among disciplines and their specialists, under such rubrics as 'cultural ecology' or 'deep ecology' (and 'environmental diversity'). At present, there is no single science of reality. The epistemological protocols are notoriously weakest for interpretation of the ontological and ethical questions that comprise the highest, or deepest, dimensions of human culture. Cultural intentions respond to always-open claims upon the possibilities (freedom) of culture and cannot be framed and made into projects for accomplishment, objectives, after the fashion of technological projects (it is precisely the defect of the post-Enlightenment utopia, or of certain totalitarian regimes, to attempt to do so).

Nonetheless, anyone involved in design must regularly make judgements which cross this divide between intentions and objectives, or, to phrase this in terms of the environmental sciences, judgements which attempt to reconcile the 'goods' of culture and of low energy use. Does this practical imagination represent a provisional fudge arising from expediency, or does it harbour a wisdom that deserves to be acknowledged? Clarification of this question will be the task of this chapter. Our guiding principle will be the notion of the 'good city', the immediate response to which would be a rich culture for the lowest energy use. However, we cannot start from scratch, nor is there available an obvious process of refinement of the given state of affairs. We seem to have great difficulty — in both architecture and in the environmental sciences — declaring positively what is this middle position to which we all profess to adhere; instead we deploy 'neither-purely-this-nor-purely-that' or 'both-and' formulations. We exist in an intermediate condition such that a cultural 'good' is obscure in its own right, as well as in its relation to technical criteria such as efficiency, by which energy use is measured.

Theory and practice: aesthetics versus technics

First, the reciprocity between theory and practice customary in the sciences and technology must be restricted to those spheres of operation, as it is not possible to specify a moral good in these terms (despite about two centuries of attempting to do just this in architectural theory). A law of science is expected to be an absolute thing (even if, within pure research, it is always held to be a hypothesis), innocent of cultural differences or concrete circumstances. By contrast, a legal law is wholly dependent upon particular people in concrete circumstances in history, both in its formulation and its interpretation (a court of law is concerned with mitigating circumstances, up to and including acquittal). Practical judgements in design are less like the application of the laws of science than the interpretation of legal laws.

Moreover, the unification of the laws of science is currently expected to be systematic in character, probably mathematical; whereas the universality of lawfulness in culture would be something like the ancient notion of cosmic justice. Since Cartesian epistemology admits only what can be known immanently, all consideration of transcendence, including that of cosmic justice, cannot attract serious attention (or is left to the speculation of philosophers, theologians and artists). Accordingly, one must return to the Platonic tradition to see a proper exploration of the possible relatedness between cosmic justice and mathematics or geometry, where, however, a clear distinction is made between these disciplines as a preparatory study to dialectic and these disciplines as deployed in, for example, the calculations of measures and quantities in building (Plato, *Republic* 510c–511e and *Philebus* 57a–59e). Nonetheless, the adherence to a completely immanent understanding has created interesting ambiguities, which testify to a smuggled, unacknowledged experience of transcendence. For example, it might be argued that the now regular application of the word 'system' to phenomena that are patently not systematic — such as politics, health, education, law, language, economics, the environment, etc. — trades a blatant untruth concerning the nature of these milieux for a term which has attained the status of cultural metaphor intimating a species of secular transcendence. In a culture for which experimental science accounts for reality as explanation — in the form of, for example, genetic codes and the laws of physics — calling one's political reality a system is a way of indicating its ultimate anonymity, its ultimate level of accountability: a purposeless universe of matter and systems, the myth of no possibility of myth.

If 'system' is meant to conform to the element of necessity in cultural conditions, it is left to cultural possibilities (freedom) to come up with a credible moral order. That is, the obligatory element in nature, understood in scientific terms, displays only efficiencies, not moralities. Here we have three choices. We could adopt the position that moralities are species of efficiency, after the fashion of the eighteenth-century physiologues and, more recently, evolutionary biology or behaviourist psychology. Despite the achievements of these disciplines, mostly in the descriptive domain, they cannot account for themselves, much less the more ambiguous or profound dimensions of human creativity. Second, we could launch a philosophical inquiry, rooted in ontology or in ethics, which, however, is methodologically beyond the environmental sciences to accept. Last, we could try to understand better the nature of reconciliation of 'goods' in practical achievements, which is what I propose to do. The practical imagination is one area where 'interdisciplinary' is not a hypothetical concept but an everyday obligation.

In architecture and urban design, however, we immediately run into the familiar problem often termed, 'the conflict between aesthetics and technics'. Whatever might emerge from a philosophical treatment of the dialogue between aesthetics and technics, in the practical arena it tends to be a conflict. What counts for 'good' aesthetically is altogether less precise, predictable and sober in its discourse than are technical 'goods'. In the name of the highest ideals of culture, apparently arbitrary configurations can be proposed which it then becomes the responsibility (even legally) of the technical consultants to 'make work', whatever 'work' means in such cases. At the same time, it is not always the case that the optimal configuration from a technical point of view is best; for example, fully exploiting passive downdraft cooling makes considerable demands upon the layout of a building. Similarly, coupling an aesthetic preference to a precise technical objective is no guarantee of success — the contemporary Chinese obsession with south-facing housing results in areas equivalent to medium-sized cities comprising endless parallel rows of people-storage, with no hint of urbanity. The degree of conflict depends upon the designer and the project; but the vast array of codes which attend construction testify as much to a distrust of the design process as to a wealth of knowledge and experience regarding best practice.

Technics generally carry the value of necessity, and aesthetics the value of freedom. The claims from freedom are the main source of the volatility of aesthetics. Modernist aesthetics has been much less good than traditional cultures at producing a coherent background urbanity; its strength lies in the very few remarkable and singular configurations, for which the cost has been vast territories of simply bad, kitsch or banal architecture. The conflict itself has been also an expensive use of energy. This economy has recently come to be measured against criteria of sustainability, which has had the effect of reducing the scope of freedom. This indirectly raises the sort of moral ambiguity with which we began; but, because aesthetic practice is so vague, and because the conflict leaves design in an obscure gap between technics and aesthetics, there is little to be gained by further analysis of the relation between technics and aesthetics. Instead we might try to look at a more direct example of moral conflict.

Few technical consultants will reject a request for assistance on the grounds that a project is morally offensive; rather they try to ameliorate the situation as best they can. There may be good grounds in sustainability to abhor a large shopping mall, for example; but, on the whole, a consultant will strive to find a way to minimise energy consumption, even if the environmental cost of transporting all the goods, the increased automobile use and the effect on an urban centre might more than cancel out any savings. It would take proposing a concentration camp to solicit outright rejection; and yet precisely this entity illuminates the more profound dimensions of our problem. This is not a critique of environmental scientists, who were, after all, among the people who first alerted us to the consequences and practice of sustainability. On the whole, they would prefer to be able to believe in the projects to which they devote their expertise. The purpose of this (hypothetical) moment of choice is to place the two 'goods' side by side, in order to open the cultural context in which such practical judgements are made.

Thinkers as different in other ways as Hannah Arendt (1958) and Michel Foucault (1994) agree that our imaginations have, as it were, regressed to being dominated by motifs of necessity. In arguing this case, Arendt analyses the centrality to political thought and practice of the economy (by definition concerned with these necessities) and of *homo laborans*. To this both she and Foucault add our obsession with, and the elaborate administration of, health. Charles Taylor (1989) would add the two-century progressive localisation of meaning in the 'inner life' of psychology; and to the consequent decline of orientation to collective or transcendent meaning, Richard Sennett (1977) would add the decline of the public realm (now deemed alienating, somewhere between enigmatic crowd-movement and danger) in favour of a progressive valorisation of private life and the domestic. The piety of thought that still attends housing conforms to this valorisation of a peaceful, private domestic life (along with the belief that it is ideally situated away from the urban centre); and perhaps the more recent fascination with the body might be taken as the logical end-point of this style of thought. Foucault's analysis of power shows that once government becomes the administration of population (Arendt's 'society'), correlating policy with statistical generalisations about welfare, 'right from the start, the state is both individualising and totalitarian' (Foucault, 1994: 325). The administration of statistically-derived norms creates conditions where the practice of good administration (welfare) and evil (a holocaust) are the same; the outcome depends upon the character of the administrators (as well as the collaboration or resistance of the people).

Sustainability and comfort

It is in this context that concepts such as physiological comfort have become the medium of exchange between efficiency and morality, between the technical and analogical dimensions of the culture. To orient environmental objectives to the norms of human physiology seems reasonable; everything outside these norms is, in theory, discomfort. Unlike moral norms, physiological norms have the character of being uncontroversial, culturally neutral (as are, unfortunately, architectural settings governed by the usual standards of comfort — by neutralising environmental difference, the image-merchants are left free to believe that the cultural aspects of an edifice are dominantly aesthetic). This notion of reasonable objective neutrality is central to our conception of design or of planning, where the priorities are expressed in terms of the economic management of labour, health, the private life — that is, the instrumentalisation of bodily necessities. These are motifs that presently enjoy the attributes of sobriety, rationality, concern, effectiveness — not to say profitability, productivity, etc.; all items on the national agenda.

The concern, sobriety and so on are genuine, and my mention of the holocaust seems at first to be indecent or hyperbolic pleading. However, for these very reasons, one is obliged to acknowledge that the camps were organised according to the objectives of factory-design, and yet it is embarrassingly obvious that the residential sectors are housing reduced to the cruellest minimum of privacy and comfort (Figure 3.1). The result in both the camps and in housing is an organised zone of accommodation, where the ground becomes an administrative planning-surface, the chief vehicle for an effective coordination of concern (attention to detail, precise mobilisation of resources, etc.) with statistical generalisations about human welfare, embodied in a topography with those characteristics (all phenomena are made to conform to the systematic coherence of the project). With ghastly irony, the residential factory in which the inhabitants are forced to produce their own death symbolises a fundamental moral ambiguity in the economic management of labour, health, the private life.

Similarly, the crisis or opportunity of sustainability presents itself as the reintegration of nature and city, a more sane dialectic of conditions and possibilities. And yet, does not its principal motif, holistic thinking, call for an even more exhaustive and thorough planning than that which we have been considering so far? Is not the famous exhortation to 'think globally, act

3.1
Hong Kong housing

locally' a motto encouraging creation of a global ant-hill of networked entities busily maximising the metabolism of physical necessities? Is not the ultimate fulfilment of the call for waste-management a perverse biotic alchemy, where people become both resource and product? Certainly the argument of Natural Capitalism (Hawken *et al.* 1999) raises such questions. Retailing their vision as 'the next industrial revolution', the authors are nonetheless concerned to be realistic, to start from the current situation. This means accepting the current forms of self-assessment and production, but including real costs in the former, and striving to make the latter as efficient as nature. The title is apt — the argument is built on a vision of nature understood in economic terms; and their picture of a reformed capitalism is a vast infrastructure — the ant-hill. Buried in the impressive wealth of practical ideas for trimming, recycling and reinventing are such questions as 'Are there ways to restructure economic activity that reward social enrichment and that reinvest in social systems' [sic] capacity to evolve ever more diverse and creative cultures?' (Hawken *et al.* 1999: 287). These latter cultures are not discussed, but are rather presented as beneficial by-products of economic behaviour. Despite the obvious good intentions, their argument simply provides one with the detailed means and good conscience for preservation of the status quo, the fanatical organisation of basic needs.

Foucault's individualisation and totalitarianism are not independent phenomena, but reciprocal aspects of the same phenomenon, which shows that the ambiguity of intentions is quite serious, and lies at the heart of our self-understanding. First, orientation to the instrumentalisation of bodily necessities effectively inverts cultural priorities: the political imagination is devoted to the amelioration of basic needs (improvement of the standard of living) rather than deploying given resources for the sake of creative politics or culture (which seemed so obvious at the beginning of this chapter). Second, this concern takes the form of reasonable objective neutrality, with the effect that the difference between the camps and housing appears to be one only of degree. The impression that the concept of comfort could mediate between the technical and analogical dimensions of the culture does not come from the environmental sciences; they are perfectly correct to believe that their procedures are only more precise forms of practice shared by the culture as a whole. However, under these conditions, comfort becomes more than a physiological norm. It aspires to be that which is common to all dimensions of the culture, and even to become a species of cultural objective (like its cognate, health), potentially taking its place alongside such concepts as peace and freedom. A technical norm becomes confused with what is common to all, on which politics or ethics are based. This scrambling of objectives and intentions happens without anyone arguing for it, least of all environmental scientists, but rather as a result of a general orientation of the culture toward lifestyle or well-being. We have arrived at a position apparently the opposite of where we began: what is missing is a substantial difference between the technical and the moral/political determinations of 'good'. Insofar as the ambiguity of intentions exposes a problem, the fault lies less with the technical imagination as such than with our collective, cultural understanding.

With regard to what is common, Aristotle, writing about politics, declared:

> The life according to the good is the greatest end both in common for all [*koine*] as well as for each separately [*choris*].
>
> (Aristotle, *Politics* 1278b, 23)

The phrase 'common for all' was, for Aristotle, neither a specifiable 'good' (whether comfort or peace), nor a mere aggregate of individual freedoms, but something in its own right — first, the *polis*, and then, more fundamentally, what was truly common to all, the transcendent conditions. This relation between conditions and possibilities came to be expressed as the reciprocity between *physis*, the nature which harboured the gods, and *nomos*, human conventions and customs. Having, therefore, the character of necessity and freedom, the (divine) lawfulness of the first was taken as a guide to the framing of (human) laws in the second. The cultural possibilities were considered a fulfilment of the given conditions; and freedom was taken for granted, as freedom-for, as the basis of one's commitment to, or solidarity with, this common-to-all.

In this understanding, there is an open reciprocity between conditions and possibilities which is made even more open in Aristotle's study of the practical life (Aristotle, *Nichomachean Ethics*, Book 6). Here we learn that there will always be a certain messiness in the practical life by comparison to the clarity with which these conditions can be apprehended in contemplation, but also that it is only in this messy, approximate milieu of practical judgements that the (universal) conditions are made concrete to (finite, particular) humans. Conversely, every involvement of a practical situation necessarily solicits a claim from the ultimate conditions, presenting the opportunity or invitation to respond with respect and creativity. That is, what was earlier called, 'the moment of choice', is part of how we interpret our freedom, our commitment, not part of the management of necessities.

Topography and 'earth'

Having declared God, or the gods, 'ridiculous' (Schiller 1982), 'an hypothesis' (Laplace 1796) or 'dead' (Nietzsche 1974), this matter of the claim of the ultimate conditions is something of a mystery to us. Rather than directly confront the problem of religion, we will take up a more concrete aspect of the issue, by looking at a term which has so far played a minor part in this discussion — topography. 'Topology' is now reserved to the mathematical sciences, and so 'topography' is preferred because it incorporates involvement with the earth. Conventionally, the earth is the subject of two reciprocal determinations. It is first a receptacle of necessity. As the biosphere, it harbours all the resources for life which we are mismanaging; and as the earth in which we build, it harbours a resistance, against which we struggle to secure foundations, for example. As 'matter', earth penetrates all physical properties of anything we make, generally manifest in the elaborate and refined calculations of the physical sciences (which also determine the earth's existence as an astronomical entity). Here again it displays the valence of resistance, as that against which scientific theories are tested. Second, the earth, as nature, is the object of sublime experiences of beauty, ranging in character from fairly kitsch sunset scenes through an affection for peasant life (deemed close to nature), to analogical dimensions of materials ('warm' wood, the iconography of marbles, the Neoplatonic conception of matter as resistant to 'idea', etc.) to the profound symbolism of the paradise garden or of the perfection of celestial movements. This last motif is very ancient, originally voiced out of the experience of nature 'red in tooth and claw' and came to signify not only a harmonious relationship with nature (and its gods), but even more, a highest truth about being at all. Since the Industrial Revolution, from which time technology is seen to progressively penetrate nature (potentially becoming a species of second nature), 'nature' has acquired the status of a standard for measuring artificiality, itself bearing the valence of 'untrue', and the object of concern, even remorse.

The potential for discourse between these two determinations of earth is, as we have seen, limited — the criteria of necessity, in the first, do not easily convert into the criteria of beauty or truth, of the second. It helps partially to recognise that they are both concepts; but this suggests only that what we are seeking here underlies the possibility for concepts as such. Staying with our effort to understand the phenomenon as it is constituted in practical life, we might think concepts have a life of their own; but, insofar as concepts are understood and put to use, they require particular people in particular circumstances in history. It is most important therefore to grasp the nature of these circumstances better; and it is perhaps surprising to recognise that what underlies and makes possible all circumstances is 'earth'. Any human situation happens somewhere, sometime. This apparent banality is a real clue to our problem. We are not disoriented temporally, in the constant exchange between memory and anticipation, neither are we without an objective experience of reality, because we have a 'there' which is always already there for everyone. This perhaps puzzling formulation accomplishes two things. First, it reverses the infamous Cartesian dictum to read, '*Sum, ergo cogito*' (loosely, 'Being precedes thinking'); and, second, it declares that that being's 'thereness' is the basis of its objectivity, its constancy and its claim upon us (Aristotle's declaration that the particular is our only concrete experience of the universal is an equivalent formulation). The part of 'thereness' with which we can be involved is 'earth'. All purely physicalist determinations of 'earth', whether technical or aesthetic, are, like any concept, made possible by this more primordial earth.

To understand this proposition concretely, we might look at the situation of a debate. Presuming it to take place in a room, it is evident that the thoughts and speeches of the participants move very rapidly and dominate attention. However, there is a more silent or implicit structure that grants the debate its freedom to be a debate. The speeches or thoughts are conditioned by postures, which are less mobile; these postures are conditioned by furniture, which is somewhat more stable; and the furniture is conditioned by the walls and floor of the room, which are most stable. The room establishes a background, a horizon for the debate, which would collapse if this implicit structure could not be taken for granted. This structure is implicit only because attention is dominated by the question of the debate; as soon as attention is devoted to this background, it ceases to be implicit, but also the question of the debate is lost. However, this shift of attention is not, as Descartes's (2001) maxim implies, a change of reality, since the 'thereness' (or 'roomness') which allowed the arguments, thoughts and words to be mobilised remains already-there for a return to the debate or another shift of attention. What lies beyond the room is even more silent or implicit, extending perhaps to a whole city, ultimately to 'earth' as that which is common to all circumstances or situations. The conditions for this debate also comprise our cultural predisposition or orientation — our prejudices, knowledge, capacity for dialogue, etc. — given spontaneously in experience (experience is all of this, not simply physical responses). The physical and cultural conditions together allow the debate to prevail; and the remoteness of the physical horizon resonates with that of the cultural conditions, such that, ultimately, 'earth' connotes their unity. When we use such expressions as a 'ground for thought' we instinctively acknowledge this structure. To this extent as well, architecture may be thought of as the preliminary horizon for praxis, which in turn depends upon the movement from primordial earth to topography, where continuity and distinction first appear. Any actual topography is a particular interpretation of the deeper principle of continuity and distinction — primordial earth is a universal only (partially) apprehended in particular circumstances. Notice, incidentally, that 'space' is unnecessary to this understanding, which speaks of a communicating structure of differences rather than the environmental simultaneity of 'space'.

If 'earth' in this sense is that which is common to all practical circumstances, it is also common to all concepts that are understood and used and therefore is that which is common to technical 'goods' and to moral or political 'goods'. This perhaps unexpected result tells us why 'topography' is so important. It is more than that which all participants in design hold in common; its proper structuring is the equivalent of a laying-out of our moral and political horizons or conditions. It is the moment of most concrete involvement with the conditions as such (Aristotle's 'universal'). As was said earlier, this presents itself as an opportunity to our freedom or commitment, and should not be governed by planning as the management of necessities. It solicits varying degrees of respect or creative interpretation according to talent, depth of consideration, and so forth.

Motifs of care, preservation or restoration of the actual, physical earth have been part of the vocabulary of concern for centuries, without, however, also being properly qualified as aspects of freedom. As part of our freedom, this insight regarding earth will always be subject to the forces of relativism, although its fundamental orientation to motifs of constancy or typicality — to what things hold in common — offers the possibility that the experiments or radical propositions might resonate with greater profundity. Correlatively, this insight is obviously not a theory which can be put into practice. It is rather an orientation or guide to thought, and to making. As such a guide, it has the merit of bringing the necessity of scientific description into closer proximity with the beauty or truth of those other forms of discourse, through orientation to that which is common to, provides the conditions for, and therefore precedes and outlasts, every particular description or evocation.

What does this give us? We have suggested that the tension between moral/political intentions and technical objectives (together comprising culture) possesses a commonality, 'earth' (therefore a general reciprocity between earth and culture). However, is not such a dimension of commonality so remote from the details of this tension as to leave it intact? There is no middle discourse which singularly articulates the neither-nor or both-and position in which we always find ourselves in the practical circumstances of design, and therefore no theory by which moral or political intentions can be converted into objectives. Rather the conflict, negotiation, collaboration within design are claimed or oriented first, by the physiognomic requirements of the architecture (architecture as preliminary horizon of praxis), then by a particular topographic order, which in turn manifests the capacity for continuity and distinction granted by earth, the ultimate horizon. As an insight which has arisen from within practice, one might offer that the ambiguity of intentions and objectives is more of an expedient fudge when it is practised as, for example, a conflict between technics and aesthetics, and more like wisdom when seen as the opportunity for better reconciling our freedom with our conditions.

Urban metabolism

This is concretely exemplified in urban design, where we are involved in the topographic structuring of the conditions for culture most comprehensively. Density as a desideratum of sustainable cities is regularly accompanied by calls for mixed use; but in practice, this has been most difficult to achieve. Instead, one most often finds mixed zoning — where monothematic zones of, for example, housing or office buildings, are linked by systems of

3.2
Shibuya, Tokyo

circulation. Zoned density is achieved through vertical packing (Figure 3.1), maximising the ratio of people to hectare, but at the cost of grim living conditions, a ground converted into a traffic and service conduit, and a net environmental deficit (even if the singularity of use allows for a significant narrowing of the environmental criteria, as we saw in the south-facing housing in China). An alternative with a better claim to mixed use is the mega-structure, such as one finds at Liverpool Street Station, London, La Defense, Paris, or, more dramatically, Shibuya, in Tokyo (Figure 3.2). Shibuya supports an occasionally delirious urban intensity; but it is the intensity of the crowd. Accordingly, it is structured according to manifold circulation routes, of which the ground level is only one and the rail and metro lines only the most mechanical. Complex spatially, economically, constructionally and environmentally, it has a global presence and has spawned continuous redevelopment in the surrounding area. Taking this larger area into account, almost every civic use is present, with no hesitation to locate art galleries several storeys underground, for example. However, even if one imagined a configuration less demanding of environmental support — and vast embodied energy — than this one, it requires a large city to thrive; and it generally has the anonymous character of a setting one passes through (at any hour) to satisfy desires or to conduct business. It is more a topography that is managed and serviced rather than a receptacle for inhabitants or other stakeholders to cultivate political self-understanding.

What both the zoned areas and the mega-structure have in common is an organisation according to systems. This conforms to the procedure of treating the earth as an administrative planning surface, and requires everyone to live systematically. The resulting rigidity with respect to change or difference means that most of urban life is sacrificed for the sake of the efficiency of planning and construction. That in fact such configurations are also rarely sustainable technically, let alone socially or politically, suggests that the emphasis upon the period of planning and construction has attained its privileged status for reasons other than either technical objectives or political intentions. Rightfully, this enigma has fallen under reassessment on grounds of long-term sustainability.

The problem of mixed use is a good example of the ambiguity of intentions. The objective of mixed use is intended to fulfil the intention of a richness of urban life. The latter is a mystery, and so mixed use becomes an equivalent that connotes the possibility of specification for design — perhaps as types or areas or circulation structures or rental profiles, etc. However, the term 'mixed' suggests an absence of structure, even chaos (at best it is the negative correlate of 'system'). Moreover, if one thinks the problem through temporally, it is evident that this is more complex than trading shoe shops for hair stylists, and may involve quite dramatic changes of use when, for example, a region of a city changes character. Unless one is to imagine wholesale reconstruction every generation or so, a certain amount of generic neutrality will be required, although it must be of a different order of flexibility than the open floor-plate of speculative office construction, or the fit-out regime of malls. What is wanted is a permanent order which is the basis of the history and identity of a locale, whose modes of change are embodied in the inherent potential for diversity.

Such a configuration has significant environmental implications: the primary environmental order must be established at the urban level, since the more precisely specified is an installed system, the less it is able to accommodate a sequence comprising doctors' surgery, restaurant, offices and dwellings, for example (a sequence derived from a large house in Spitalfields, London). In general, this places a priority on natural ventilation and lighting, and additionally on being able to create real environmental difference (between street and park, for example); but, if one is also to imagine the diversity to manifest itself in discrete addresses, the integrity of these clusters must be preserved without compromising the whole. That is to say, there is to the topography of structured diversity of use a commensurate environmental topography, which incorporates an inherent dimension of mutual responsibility of the parts to the whole. Phrasing these desiderata as a matter of practical realities, it appears that the objective of mixed use as a plausible receptacle for culture (socially, politically, environmentally) is best understood as a topographic order which supports the structured mediation of conflict. Because, in practical terms, this involves qualitative judgements regarding a rich ensemble of cultural and topographic processes that encompasses an extreme range of temporalities, from centuries to hours, the term 'urban metabolism' has evolved to describe the phenomenon which is the topic of both design and understanding.

As it happens, we have abundant examples of such topographies in the large-block metabolism of most European cities, whose qualities at

3.3
Large block structure, Padova, Italy

least establish a standard by which to measure any alternative (Figure 3.3). The potential heuristic value of these configurations is often dismissed on the grounds that their richness is the result of historical accumulation, and it is no good trying to simulate this in design. This utterly misses the point, and betrays a fatal adherence to the view that the long-term life of a city should be measured against what can be made an objective for the period of planning and construction. The point is that these configurations have been able to sustain themselves across centuries of change — from the horse, through steam and electricity to automobiles and computers — even under the restrictions of (very recent) preservation orders, which mostly affect the exteriors. The actual structure of change in these configurations exhibits several layers. The perimeter is most stable, usually lasting for several centuries. Within that, one sees transformations in temporal cycles (and in scale) ranging from major transformations (approximately every century) to lesser interventions (every 50–25 years) down to refurbishment and fit-out cycles (10–5 years) as well as seasonal and daily modifications.

These configurations are sufficiently rich in their metabolism to withstand the insults and compliments of generations of architects, not to say an equal number of ad hoc or cowboy constructions (unlike the Italian *isolati* or the IBA blocks in Berlin, they are not systematically coherent; the urban and the architectural orders are an open dialogue rather than a single configuration). That is to say they are robust because of the structure of richness. The criteria are relatively few, the variations endless (Figure 3.4). By large, one means something at least 200–300 m on a side. Such a block can often exhibit a very different character from one side to another (for example, one face might address a major boulevard whilst the opposite face could be more intimate in scale). A strong — usually high-rent — presence to the street preserves the — usually low-rent — integrity of the interior. The interior can contain anything from civic institutions in a park (Paris) to converted ateliers (Hackesche Höfe, Berlin) to the town bus depot beneath ateliers and flats (Como). However, the interior

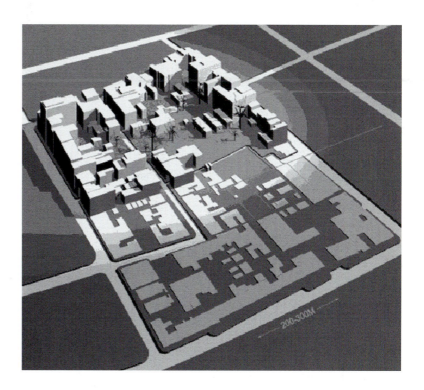

3.4
Large block, generic diagram

is rarely a single entity, but often exhibits the greatest richness or diversity, from fairly circumstantial, aggregative episodes to quite tightly-structured small towns (the *cités* of Paris), able to accommodate anything from clinics to flats to illegal activities. Sometimes parts or all of the interior are open to the street, sometimes not. The crucial factor which prevents this interior development from becoming merely a collection of light-wells and drains is an attentiveness to the stratification from the street to the interior of the block through sequences of courtyards, small streets or gardens. This not only mediates the hierarchy of public and semi-public life but does so in a manner that allows the block to breathe easily. Ideally, for these purposes (as well as for controlling run-off, for bioremediation and for exploiting the possibilities of kitchen and medicinal planting), a substantial portion of the interior is garden; but, even without this, the 24-hour life of such blocks is congenial to, for example, CHP (combined heat and power) units and the clustering to discrete, proprietary environmental modulation.

These large — or deep — blocks are always public before they are private; they command a collective order at their own scale even if one cannot get past any of the gates or doors without invitation. The richness (diversity) of urban life they support is an important contribution to their long-term stability as well as offering a nurturing context for other such blocks. Similarly, the required generic neutrality does not have the abstract emptiness of office floor-plate, but declares an institutional typicality that we glimpsed in the large house in Spitalfields. By 'institutional typicality' is meant the manner in which the hierarchy of room sizes and arrangement of such houses mediates a movement from reception to private activity which is easily adapted for offices, clinics, research agencies, even schools of architecture. To this extent the clusters always manifest the conditions of their potential diversity, which in turn qualifies the direction of change.

Conclusion

Urban metabolism teaches us that diversity or richness is itself a type of order, but one that will always elude complete theoretical description. A norm which can be technically specified has an important heuristic value, but cannot be confused with that which is common-to-all, whose determination is a matter of cultural insight and commitment. The phrase used earlier, 'a topographic order which supports the structured mediation of conflict', deserves emphasis: it literally gives the true nature of our problem. Because one half of the ambiguity of intentions (that pertaining to morals or politics) can never be placed before us and made into an objective, the ambiguity is not an issue to be fully illuminated intellectually or methodologically. It actually speaks to us most clearly as it constitutes itself within practical understanding, as a plausible configuration. With regard to the full scope of issues encompassed by urban metabolism, there is no higher level of abstraction — the difficulties of simulation are legion — than an actual urban topography, which alone claims all participants (designers and inhabitants) equally and with a precision appropriate to the case. Indeed, properly understood and from the most universal level of understanding down to the details, the problem of the negotiation between intentions and objectives looks most like one of these blocks, and particularly the manner in which they preserve and articulate the claims of 'earth' as the conditions for freedom.

References

Arendt, H. (1958) *The Human Condition*, Chicago: University of Chicago Press.

Aristotle, *Politics* 1278b, 23.

Aristotle, *Nichomachean Ethics*, Book 6.

Descartes, R. (2001) *Principles of Philosophy*, Blue Unicorn Editions.

Foucault, M. (1994) 'Omnes et Singulatum', in Faubion, J. D. (ed.) *Power: Essential works of Foucault 1954–1984*, Vol. 3, London: Penguin.

Hawken, P., Lovins, A. B. and Lovins, L. H. (1999) *Natural Capitalism: The next industrial revolution*, London: Earthscan.

Laplace, P. S. (1796) *Exposition du Systeme du Monde*, Paris: Imprimerie du Cercle Social.

Nietzsche, F. W. (1974) *The Gay Science*, trans. W. Kaufmann, London: Vintage Books.

Plato, *Republic* 510c–511e.

Plato, *Philebus* 57a–59e.

Schiller, J. C. H. (1982) *On the Aesthetic Education of Man, in a Series of Letters*, Oxford: Oxford University Press.

Sennett, R. (1977) *The Fall of Public Man*, London: Random House.

Taylor, C. (1989) *Sources of the Self: The making of the modern identity*, Cambridge: Cambridge University Press.

Chapter 4

Human nature

Nick Baker

Life outdoors

Although we spend most of our time indoors, we are really outdoor animals. The forces that have selected the genes of contemporary man are found outdoors in the plains, forests and mountains, not in air-conditioned bedrooms and at ergonomically designed workstations. Fifteen generations ago, a period of little consequence in evolutionary terms, most of our ancestors would spend the majority of their waking hours outdoors, and buildings would primarily provide only shelter and security during the hours of darkness. Even when inside, the relatively poor performance of the building meant that the indoor conditions closely tracked the outdoor environment.

It is proposed here then, that our evolved but anachronistic responses to the outdoors have an influence on how we respond to the modern indoor environment, although they are overlaid and masked by much faster moving cultural values. If this were true, what are the implications?

The most essential characteristic of the outdoor environment is its diversity. There is diversity at different spatial scales, ranging from human scale to global scale, and on different time-scales: daily and annual cycles as well as the quasi-random nature of weather. Yet it is the elimination of this diversity that is the main objective of the engineering services of a building.

A good example is the control of temperature — now one of the most ubiquitous standards in the built environment. Huge industries and professions are based upon the mechanics of providing optimum thermal conditions indoors. However, later we shall be reviewing a substantial body of evidence which suggests that people are quite satisfied with a considerable diversity of thermal conditions, and indeed may prefer it to uniformity. There is similar evidence of preferences for non-uniformity in illumination. These two parameters are not, of course, the only ones referred to by the term 'environmental diversity' but due to the ease with which they can be quantified and at the same time, invoke subjective responses, they provide some of the most interesting data.

The move away from nature

Primitive man lived almost entirely outdoors. The earliest traces of man are found in areas where the climate was (although may not be now) conducive to survival with a minimum of climatic moderation. Having no formal built shelter, nor clothing, man used his adaptive skills to find places away from the scorching midday sun and the chills of the clear night sky. In cooler climates he would often utilise the thermal stability of the ground by creeping into caves during times of temperature extremes. But by and large he was an outdoor naked ape, with sufficient adaptive skills to survive the climatic variation of the region he inhabited.

Three and a half million years later (but pre-industrial revolution) most of our ancestors still spent their lives outdoors. The commonest employment was agriculture which demanded an intimate knowledge of the climate — rainfall, frosts, wind and their interaction with the landscape — shelter, drainage, pests, etc. constantly reinforced man's link with nature, and often had a vital role in survival. Many other activities such as house-building, transportation, fishing, hunting, woodcutting, etc. also took place outdoors, and were thus influenced by weather and the hours of daylight. Even the limited manufacturing activities such as carpentry, pottery, basketry, which today happen primarily indoors, would then have been carried out in the semi-open, being totally dependent on natural light.

A few particular activities moved indoors. Spinning and weaving, the materials and equipment easily damaged by the weather, would take place inside. So too did the rare activity of scholarly study. Both of these required good daylight and it is in spinning galleries and libraries that we first see conscious efforts being made to ensure sufficient levels of indoor natural light. In providing daylight, usually through vertical apertures, a strong visual link with the outside was maintained. This was further strengthened by the fact that until well into the sixteenth century, most of the windows would be unglazed. This, together with ventilation requirements for un-flued fires, meant that the thermal conditions indoors closely tracked those outside.

In coastal regions, much of the population was engaged in maritime activities — not only fishing, but transport as well. For the mariners, the majority relying upon sail until a century ago, a thorough knowledge of the workings of the wind and tide was essential. The decision whether to stay in port and let the easterly gale blow itself out, or to keep to schedule and ensure a cargo on the return journey, was not only a matter of economic importance, but actually a matter of life and death.

The great migration

With the Industrial Revolution came the great migration. The newly emerging industries provided both the demand and the means whereby people would, increasingly, live and work indoors — first factories, then offices, shops, schools and hospitals. Urbanisation provided the workforce to work in the factories, poaching people from the land. As industrialisation progressed, so farming became more efficient and less labour intensive.

Not only was the urban building itself providing an alternative to the outdoor life but the density of development meant that 'the outside' was becoming more and more distinct from nature. Planning laws did little to control density and although philanthropic bodies and individuals were including parks and squares in their development plans, they tended to be in the wealthy areas. For the majority of urban dwellers, contact with nature was becoming a special event rather than the norm. This was not helped by the poor quality of the urban environment. The pollution emanating from domestic chimneys together with the industrial smokestacks is unimaginable to us today. As the nineteenth-century poet Theodore Watts-Dunton observed 'We looked o'er London, where men wither and choke, roofed in, poor souls, renouncing stars and skies ...'. This pollution led to a change of urban climate, shutting out the sun and precipitating polluted and poisonous fogs and rain. An indication of how significant this effect was is that, prior to the Clean Air Act of 1952, the average hours of sunshine for the winter months of November to February inclusive in central London was little more than half that of the surrounding countryside (1.1 hours per day compared with 2.0 hours per day).

In spite of this transformation of the outdoor world for the nineteenth-century urban dweller — to a landscape of sooty cliffs of brick and stone, a ground of paving, tarmac and detritus, a fetid sunless atmosphere of fog and smoke, it was still 'the outside'. Public transport exposed people to wind, rain, cold and, more welcome in our temperate climate, occasional warm sunshine. The cycles of the day and night, and the four seasons were sensed by everyone. Urban children tended to spend much of their time outdoors, probably as much as their country counterparts, simply because there was little for them to do indoors — no video, computers, centrally-heated bedrooms, etc.

In contrast, it is now quite possible, indeed commonplace, to go from a centrally-heated apartment, directly into a heated or cooled car, drive to the underground car park beneath the office block, shopping arcade or leisure centre without sensing the outdoor environment at all. Public transport is also striving to insulate travellers wherever possible, integrating with walkways, subways and malls. Few urban children walk to school and the streets, and even parks, are often considered far too dangerous places for play.

Emerging technologies

The means to support indoor living were provided by emerging technologies — iron provided large-span structures and brick production benefited from economies of scale. The industrial manufacture of glass, advanced by processes such as cylindrical glass and finally float glass, resulted in vastly reduced cost. Cast iron was used to manufacture fireplaces, stoves and radiators improving the application, control and hygiene of heating. A succession of improvements were made in the production of artificial light — the candle with its plaited wick, the Argand burner for the oil lamp, and the incandescent mantle for the gas lamp, all represented steady progress in the efficiency of light production. By the end of the nineteenth century, it was entirely feasible to use artificial light to extend the working day. With the rapid growth of electricity, the stage was set for the next step — the use of artificial light in the daytime to illuminate parts of large buildings which never received daylight.

Just as electric lighting provided independence from daylight, the development of the electric motor provided the essential motive power for mechanical ventilation, thereby no longer requiring close proximity to an openable window. By the beginning of the twentieth century the dream of a perfectly controlled artificial environment could become a reality.

Another technological development which supported man's move indoors, particularly in warmer climates, was refrigeration — or as it is usually referred to — air-conditioning. Linking the electric motor, the compressor and the heat-exchanger, Carrier the inventor spawned an industry second only to that of the automobile. Initially, the purpose of air-conditioning was to counteract the negative results of the climate — too hot, too humid or too cold. Its primary objective was not to isolate the occupant from nature. Just as the potential for artificial light to support deep plan spaces would soon be realised, so too would the removal of large quantities of heat gain from deep plan spaces become possible.

However, there was still one obstacle, and this was the cost of electricity in relation to the efficiency of the mechanical equipment, and in particular the light sources. For example, a typical tungsten filament lamp in a typical luminaire from the 1920s would deliver about 5 lumens per watt on the workplane. Compare this with a modern high efficiency fluorescent lamp and luminaire which can deliver about 50 lumens per watt. Furthermore, the cost of electricity (allowing for inflation) has reduced twentyfold, which means that a lumen of light in 1920 cost about 200 times more than at present. Thus, although it was possible to do without daylighting, there was no economic incentive — windows and daylighting made good economic sense. Finally, the greater heat production from the less efficient light sources, meant that even if the cost could be borne, there would be considerable risk of overheating, or it would make heavy demands on mechanical ventilation.

However, with centralisation of electricity generation, the cost of electricity dropped and by the post-war years with the promise of cheap nuclear power, the fully air-conditioned building began to become a reality. Particularly in the USA, the air-conditioning industry embarked on a massive advertising campaign. A number of arguments were used including productivity, hygiene, the control of nature and status. By the 1950s, air-conditioning was widespread in non-domestic buildings and a determined campaign was underway to extend this to the home, even using the economic performance of the husband as incentive to provide the 'perfect environment where housework would become a joy rather than a burden'.

By the 1970s, nature had been conquered. In the UK, the Electricity Council was promoting all-electric buildings as providing an optimised artificial environment in deep plan buildings. Narrow slit windows where the only concession to a link with nature, providing minimal but poignant reminders of the world outside. It is interesting to speculate if this attitude was also driven by the deep-seated desire to respond to, and in this case retreat from, the natural world — if so, we might say that it had become a victim of its own success.

Our special relationship with nature

> Few people have a problem with the idea that humans are descended from apes. But while people believe that our general shape and structure are derived from other creatures, few consider the psychological implications. Man not only looks, moves and breathes like an ape, he also thinks like one. It is back in our primeval past that we find the first clues to understand our human instincts.
>
> (Winston 2003: 3)

The conjecture outlined in the previous section, that our evolutionary background is equally important to our response to the built environment as many of our social attitudes, is attractive; but where is the evidence? In the following section, we outline a number of studies which, although not specifically designed to demonstrate this significance, provide results that can be interpreted to support it.

Thermal comfort

Thermal comfort has one of the strongest and most obvious links with survival. As warm-blooded animals, we have to maintain our core temperature at 37°C +/- about 1°C from birth to death, in environments ranging from tropical to polar. This is no easy task and failure to do this has dire results. It is as serious as inability to find food, or from the viewpoint of the species, failure to find a mate. In response to this, man has developed a wide range of actions — from creeping inside a cave to the air-conditioning of buildings.

In our contemporary life the notion of comfort is not seen to relate closely to survival. Rather, it is something between an expectation and an indulgence. Feeling a little chilly and deciding to take one's gin and tonic in from the patio to the warmer lounge, is hardly likely to be seen as an act of survival. But, for certain, the mechanism of thermal comfort is just the same as that which, in the natural world, enabled us to meet the thermal challenge. If we are looking for evidence of vestigial responses to the natural world, it seems appropriate then, that our first area of interest should be thermal comfort.

Thermal comfort theory

More than 20 years ago, Heschong (1979), in her highly original book *Thermal Delight in Architecture*, decried thermal uniformity. A decade later, this time as a result of rigorous field studies, Schiller (1990) concluded that 'even people voting with extreme [thermal] sensations are not necessarily dissatisfied'. Since then many field studies have confirmed that thermal diversity is tolerated, and in many cases enjoyed.

Why then, does the notion of optimal thermal conditions still exist? Surely it is because optimised conditions can be predicted by a physical thermodynamic model. The model may not be simple in structure, but the concept is simple in that it obeys reliable laws of physics, occupying a single domain, and uncomplicated by behavioural and psychological factors.

This approach, most developed in the heat balance model by Fanger (1970), relates thermal sensation to the imbalance between heat generated by metabolic activity and the heat lost from the body to the environment. Written mathematically it is represented in empirical equation:

$$PMV = 0.303 \, (e^{-0.036M} + 0.28) \, (M - H) \qquad (4.1)$$

where:

M is the metabolic rate in watts
H is the total heat loss from the body to the environment
PMV is the predicted mean vote, a five-point scale from cold to hot.

It can be seen that when the heat loss H is equal to the metabolic heat production M, the difference being zero results in a Predicted Mean Vote (PMV) of zero, indicating neutrality. Any departure from equilibrium, irrespective of time-scale, will predict a non-neutral sensation.

Fanger goes on to relate the thermal sensation as indicated by PMV, to a value judgement of this sensation, the Predicted Percentage Dissatisfied (PPD), shown in equation (4.2) and Figure 4.1. It indicates that non-neutral thermal sensations always tend to dissatisfaction and it follows from the first equation that thermal imbalance must also always lead to dissatisfaction. Since, for a given metabolic rate and clothing insulation level, the heat balance is primarily controlled by the environmental temperature, it is easy to see where the idea of the 'universal and optimal temperature' came from.

$$PPD = 100 - 95 \, \exp[-(0.34PMV^4 + 0.22PMV^2)] \qquad (4.2)$$

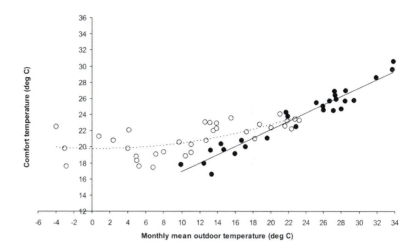

4.1

Scatter diagram of neutral temperature showing points and regression lines for free running buildings (continuous line and black circles) and other buildings (dashed line and white circles) (Humphreys 1978)

It follows from the concept of optimal temperature that both spatial and temporal deviation away from the optimum value must in some way degrade the quality of the environment. Although some degree of tolerance is accepted, close control remains the target of conventional services design.

In the early 1970s, it was pointed out by Humphries and Nicol (1970) that no such universal optimum temperature existed. Figure 4.2 shows the much quoted result of a survey of surveys, relating preferred comfort temperatures to mean monthly outdoor temperature for comfort surveys conducted around the world. It shows that for naturally ventilated buildings, the preferred indoor temperature responds to the outdoor temperature. Only in air-conditioned buildings is there any evidence of a universal optimum, and thus a self-fulfilling prophecy.

Much work has been carried out since then (e.g. Brager and de Dear 2000) which demonstrates beyond doubt that people are satisfied with a much greater range of thermal conditions than Fanger's theory predicts. How can we explain the paradox? Provided the body being

4.2

Relationship between the predicted mean vote (PMV) and the percentage of people dissatisfied (PPD) (Fanger 1970)

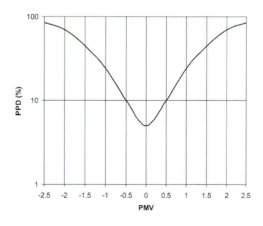

described is in a steady state, the physics on which the heat balance is based is not in dispute. However, although the body is held to very close temperature limits 37°C +/- 1°C, it has a substantial thermal inertia. For example, an imbalance of 100% of the metabolic heat, say 140 watts, would have to exist for 40 minutes to raise the mean body temperature of an adult 1 K. Thus, describing the comfort mechanism using a steady state model only is misleading. In fact the body is in a dynamic state, and what matters for health, and ultimately survival, is a longer-term heat balance, and the limitation of temperature swings within limits.

The second omission is not in the physics, but in the psychological model — the correlation between thermal sensation and dissatisfaction. This correlation was obtained from laboratory conditions where all parameters relevant to thermal balance were strictly controlled. When experiencing non-neutral conditions, no actions by the subjects to mitigate discomfort were permitted. This is significantly different from real contexts where, with very few exceptions, some opportunity to make adaptive responses always exists. These could include adjustment to clothing or posture, moving to a warmer or cooler part of the room, or making some adjustment to the room itself such as opening a window. The omission of this behavioural component from the model is crucial. There is now a growing body of evidence to show that the existence of adaptive opportunity greatly extends the acceptability of an environment (e.g. Guedes 2000). Furthermore, there is some evidence that the presence of stimuli to take adaptive action is a positive, rather than negative, attribute. This could be explained by our evolved response to the natural environment, which shows a high degree of spatial and temporal variance, requiring us to exercise our adaptive skills. We will return to this issue later. This can certainly explain the choices we make for recreation and leisure, where not only do we routinely expose ourselves to thermal extremes, but also other physical discomforts and dangers.

Field studies

de Dear (1998) lists numerous international surveys which show the effect of adaptive actions in significantly increasing the tolerance of subjects in real contexts. Two specific examples are described here which illustrate the kind of evidence. In a survey in Athens carried out for the EU Project PASCOOL (Baker and Standeven 1994), seven workers in an office were monitored in detail for several weeks. During this time they made 345 adaptive actions (clothing adjustments, window opening, deployment of blinds, moving to cooler part of the room, etc.) and recorded a preferred temperature of 29.5°C, about 5 K higher than a classically predicted comfort temperature (Baker and Standeven 1996). A second, much wider study (Guedes 2000) assessed the comfort of 560 workers in 26 office buildings in Lisbon. Amongst much other data a surprising finding was that in heat stressed circumstances subjects showed a significantly higher level of satisfaction where they perceived a

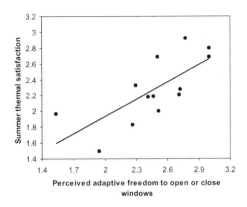

Perceived adaptive freedom of opening or closing of windows during summer monitoring in Portugal, where each plot represents one building (Guedes, 2000)

freedom to take adaptive actions, even if they did not in fact take them (Figure 4.3). This latter data demonstrates a strong psychological factor

A study on outdoor comfort by Nikolopoulou (1998), described later in this book, revealed a dramatic increase in tolerance of 'non-optimum' thermal conditions, compared with theoretical predictions based on the heat balance model, and far more tolerance when compared with actual comfort votes from people indoors. Futhermore, the subjects showed greater satisfaction with the prevailing conditions when they were outdoors by free choice, rather than when waiting to meet someone.

This evidence suggests that humans have a special empathy with stimuli which can be seen to be from natural causes, and that this is even stronger when the individual has freedom to make a response, such as moving from the shade into the spring sunshine.

Daylight and views of outside

Another study carried out in Cambridge by Papairi (1999), also described in detail later in this book, was concerned with visual comfort. She compared actual responses of students working in four libraries with a conventional assessment of visual working conditions measuring illuminance level, glare, etc. The results were initially surprising; the library where bright sunlight and dark shadows fell on the work area was preferred over the visual calm of a study carrel illuminated with glare-free diffuse daylight from light-reflective surfaces — conditions which conventional theory would have expected to score highly. However, the characteristics of the sunlit library were, first, that it had fine views across the river Cam and, second, that the occupant had the option to turn away from the sunlight or move deeper into the room to avoid it altogether. As in the previous cases reported above, the presence of adaptive opportunity seems to be a key parameter, together with the association of the stimuli with 'natural' causes.

There is a large body of work suggesting that daylight itself has an intrinsic value. This ranges from the need for daylight to achieve the synchrony of the circadian rhythm — the absence of this synchronising effect causing

Seasonal Affective Disorder, or SAD (Cawthorne 1994) — to the improved performance of schoolchildren in daylit classrooms compared with those under artificial lighting (Edwards 2003). In the former case, the key parameters are the intensity and the time of day that the light is delivered, whilst in the latter case the spectral content of light seems to be most significant.

The value of daylight and sunlight in hospitals has long been recognised. In a survey of 30 European hospitals, Saxon Snell (1883) makes a typical observation of the General Hospital of Berlin:

> The free admission of sun, light and air, to every part of the sick wards, and the regulation of ventilation by natural means were the subjects of minute study and attention.

In a comparative analysis of the 30 hospitals he includes tables of the area of glazing and openable area of window per bedspace. These values ranged from 1.5 to 3.5 m^2 and 0.5 to 3 m^2 respectively. The role that sunlight was expected to play was not explicitly mentioned — it was certainly associated in part with its sterilising role due to the UV content — but it seems likely that its psychological value was also recognised.

Even without view, the dynamic quality of daylight seems to have an intrinsic value in the healing process. Keep *et al.* (1980) report on a comparison between the Intensive Care Units at Plymouth and Norwich. It was found that patients from the windowless Norwich unit had a much less accurate memory of their length of stay and were subject to greater problems of disorientation. The incidence of hallucinations and delusions at Norwich was twice that at Plymouth, although the windows at Plymouth were only translucent, providing daylight but no external view. From the library study mentioned above it is clear that the value of daylight is at least in part related to the presence of an outdoor view. This has attracted attention in healthcare. Work carried out in the 1980s by Ulrich (1984) showed that patients recovered more rapidly and using fewer painkillers when able to view a middle-distance natural scene including trees, than when viewing a blank wall (Table 4.1).

Before leaving the topic of daylight, it is appropriate to report one more familiar phenomenon which seems to integrate many of these objective and subjective effects. It can readily be observed that people sitting by a

Table 4.1 **Analgesic doses per patient of 46 patients between 2 and 5 days after surgery, comparing patient groups that have a view of nature ('tree group') to those that do not ('wall group') (Ulrich 1984)**

Analgesic strength	Number of doses	
	Wall group	Tree group
Strong	2.48	0.96
Moderate	3.65	1.74
Weak	2.57	5.39

window will find the diminishing illuminance level as twilight approaches quite acceptable, even when engaged in visual tasks such as reading. They will tolerate illuminance levels which if provided from artificial sources would be regarded as quite unacceptable and would contravene all functional standards.

Nature in buildings

Man has an almost universal love affair with gardens. Our gardening endeavours range from the humblest window box to massively expensive planting schemes in public places, or where historically the value of the planting has been accorded almost a divine status, such as in the Moorish gardens of the Alhambra (Figure 4.4). Even the most formal of gardens, in spite of their high level of human intervention, allow nature in: the dependence upon the seasons; the day-to-day weather; and the vagaries and uncertainties of the botanical occupants contrasting with the permanence and artifice of the infrastructure. At the other end of the scale, the English tradition of landscape gardening is perhaps a poignant indication of man's love of the wild, albeit often an idealised and indulgent one.

4.4
View of a garden at the Alhambra, Granada, Spain

Table 4.2   **Effect of indoor plants on neuro-psychological symptoms (Fjeld *et al.* 1998). Scoring: no problem — 0, minor problem — 1, moderate problem — 2, major problem — 3**

Symptoms	No plants	With plants
Fatigue	0.82	0.58
Heavy-headedness	0.71	0.58
Headache	0.33	0.27
Dizziness/nausea	0.27	0.22
Poor concentration	0.50	0.42
Mean sum score	2.6	2.0

As man has retreated further into the indoors, so he has brought tamed nature, the garden, indoors with him. Conservatories and atria are large-scale manifestations of this. Ideally, the atrium would be so designed to give the plants sufficient daylight and fresh air to thrive. All too often, however, the functional design of the atrium becomes subordinate to an architectural programme, and the plants, like the building's occupants, have to be sustained by energy-consuming services.

At a smaller scale, we see the ubiquitous use of indoor planting. The plants, usually tropical forest species, do not enjoy their habitat, but the occupants enjoy them and even seem to show measurable benefit from them. In a study in Norway, Fjeld *et al.* (1998) compared two large groups of office workers in a cross-over study in about 60 individually occupied office rooms, one group having quite extensive foliage planting, the other group having none, alternating over random periods. Recovery was faster and complaints of neuro-psychological symptoms, such as fatigue, headache and concentration problems were reduced by 23% in the case of the rooms with plants, similar to the reduction in mucous membrane problems (Table 4.2).

In earlier work (Wolverton *et al.* 1985) there had been reports of physical improvement in air quality by indoor planting, namely the reduction of contaminants such as formaldehyde, benzine and carbon monoxide. However, in the study by Fjeld *et al.*, due to the nature of the filtered mechanical ventilation system, it was claimed that the air quality was already high. In conclusion, the authors suggested that 'an improvement in the feeling of well-being' was the most likely explanation for the reduction in the complaints — by implication a psychological effect due to the visual presence of vegetation in the room.

The evidence reviewed above suggests that the psychological effects are not entirely separate from the physical effects, and that there may be some interaction, even across domains such as the thermal, visual and air quality. Could it be that it is the holistic effect which engenders a special quality — the true quality of natural ambience? For example, we describe a building as 'naturally ventilated' if unprocessed outdoor air enters the building under natural forces. Usually this takes place through an open window, but not necessarily. If the fresh air entered through a grille mounted in the ceiling and was controlled by a remote infrared handset, would the satisfaction be the same as walking over to a window and opening it — getting glimpses of the outside and the sudden audibility and odour of outdoor life (be it good or bad)? And does the intuitive means of connecting with the exterior, already visible through the window, engender a special 'added value' to the action of opening the window? Similar questions could be asked about advanced daylighting systems — light ducts, fibre optics, etc. — where the entry point of the daylight to the room is indistinguishable from an artificial light source.

It may be then that it is the whole package — the multiplicity of mutually associated stimuli together with the effectiveness of an intuitive response — that is necessary for the full benefit of the environmental diversity to be gained.

Implications for architectural design

We have suggested that man has a need for environmental stimuli and a need to respond to them. If this is true, what are the implications for the design of buildings? We have also implied that these stimuli should be due to natural causes and associated with the natural environment (although this could be simply because the positive evidence available is only from cases where the stimuli are of that type). We refer to this package of stimuli as 'ambience'. This prompts the following questions:

- Is it essential to have *natural ambience* by contact with natural environmental diversity?

or:

- Can we create *artificial ambience* — where natural environmental diversity is simulated?

or even:

- Can we create *synthetic ambience* — were the diversity is artificial *and* arbitrary?

Natural ambience
Creating a natural ambiance by direct contact with the natural environment is the conventional view. The architectural interpretation is the adoption of shallow plan buildings, naturally ventilated and daylit with openable windows. Controls would be intuitive and sympathetic to occupant participation, and

the spatial and technical design would provide diversity and adaptive opportunity. Intermediate spaces such as atria, conservatories, loggia and verandahs, free from active control, form a soft edge between the interior and exterior. Externally, the architecture continues into the garden or park where the microclimate still provides a level of moderation and the horticulture is applied with a degree of artifice, but ultimately allows nature to dominate the environment. The landscape design is influenced by the perception of the occupants in the building, rather than being seen as a setting for the building seen from outside. The principle continues at the urban scale, with accessibility of and to wildlife considered in the provision of green corridors and wild parks.

It is a person-centred architecture where the context is the natural world and the building is seen only as a mediator. The contextual awareness does not stop at the site boundary: it is reflected in a concern for the global environment through the choice of materials and a responsible attitude to the use of energy and other resources.

The above agenda would hardly seem controversial to many architects even if resources were plentiful. It was a principle followed in the design of many of the great houses and gardens in England of the eighteenth and nineteenth centuries. It would not be out of place in a brief for a modern company headquarters set in a business park, although it may require an enlightened client to resist calls from the building economists for restraint.

Why then, do we have to consider the issue further? Urban growth, the coalescing of communities, seems to be driven by a force as inevitable as the law of gravity. Unlike gravity, it is not described by a simple algorithm — rather it is the result of complex political, cultural, functional and environmental expedients, which will not be discussed here. The outcome, however, is relevant, since together with the resulting growth of land value, it has led to an ever-increasing size of building and plan depth. This in itself removes people from the natural ambience of outside. Furthermore, due to the density of building on the ground, the external environment with its noise, pollution and hard surfaces further alienates the urban dweller from nature.

Just as at the end of the nineteenth century the developing technologies acted as a stimulus to urbanisation and the interiorisation of the working environment, current technologies offer opportunities that creative architects find irresistible. Recent developments in materials such as, for example, glass, polymers, stainless steel, in computed structural analysis, and in information technology, all facilitate the increase in size and technological complexity of the modern building.

Inevitably then, the question must arise — can we do without natural ambience? Can environmental diversity be delivered in a different way?

Artificial ambience

The notion of recreating environmental diversity artificially is nothing new. In a less technological age, evocation of the outdoors was provided by painting and sculpture, spanning perceived levels of taste from fine art to the 'high-naff' of plastic flowers (with perfume!) and animated pictures of waterfalls. With current information technology it would not be difficult to offer a rich menu of naturalistic stimuli — images of landscape and its inhabitants, sounds and even smells could be delivered deep into a building. This could transport the occupant to distant idyllic environs, or simply relay the real outdoor surroundings of the building. It could be accompanied by naturalistic environmental stimuli such as temperature swings and modulations of luminance and colour temperature.

Simulation and virtual reality has reached an advanced state of development — now used for applications as diverse as, for example, training in surgery, flying and presenting building 'walk throughs' from electronic moving images developed straight from CAD packages. Simulation in these circumstances is hugely successful and convincing — it is well known that airline pilots training to cope with emergencies show signs of profound stress although they are quite aware that the circumstances are not real. If this is so successful, would not the evocation of the garden outside be an easy task?

In the case of the flight simulator, the illusion is the focus of interest. Having accepted the illusion, the subject is sympathetic and voluntarily carried along by the illusion. In contrast, the image of distant mountains projected onto the wall of a building is meant to be absorbed subliminally if it is to achieve the quality of natural ambience. This is fundamentally different from the flight simulator pilot, and unlikely to be convincing. However, this is an unresearched area and it would be wrong to abandon the solution without test.

Synthetic ambience

We have made the case for the role of environmental diversity in stimulating adaptive behaviour. But does this diversity have to relate, either directly or by artificial means, to nature? Could not the thermal, visual and acoustic environment be modulated in an arbitrary way, and a new set of adaptive opportunities be created artificially? For example a temperature swing could be delivered by the air-conditioning system at the same time that a strong visual event was created by the lighting system. This could then be neutralised by an action through a graphic interface on the occupant's workstation. Would this be as satisfying as walking to the window and throwing it open? It is difficult to respond to such a question other than negatively — but is this just a sentimental response? Again, it is an unresearched area.

A view of the future

Most people would probably favour the first option — the natural ambience. What future does this have?

Today, from a fifth floor apartment in the centre of downtown Athens, I look across a narrow but busy street onto a balcony. It is laden with many pots containing green plants, vines and flowering shrubs of several varieties. A lady comes out and after some time sweeping, tidying and watering the plants from a can, she returns from indoors with a songbird in a cage which she hangs from the canopy framework. Moments later she is joined by another lady with two cups of coffee. They sit down, arranging their canvas chairs so that they catch the winter sunshine as it slants under the canvas canopy. Meanwhile, five storeys down, the traffic growls its way through the polluted street.

This encouraging image gives hope that, when given the opportunity, people will green their cities. In several countries, there is now serious consideration to positively encourage wildlife back into the city — providing green links and bridges, roosting places for birds, etc. Surprising benefits often result — the undesirable population explosions of starlings and pigeons being controlled by establishing populations of birds of prey, such as the kestrel and the peregrine falcon.

There are few cities, however historic and full of cultural delights, that do not experience the weekend exodus of people travelling to the country. For the rich, this will be to their second home, for the less well off, a visit to the forest, the hills or the coast. This pattern is repeated at a larger scale with air travel, providing weekend breaks to other countries and climates, at huge environmental cost. If urban design could include sufficient access to nature within a local area, there could be massive savings in transport energy.

Conclusions

We have made a strong case for environmental diversity. Is it justified, and if so how is it to be delivered? If we accept that there is a case, there seem to be two possible directions:

1   embrace 'real nature' (naturally ventilated, daylit buildings, with user-controls, set in an accessible, naturalised landscape into which nature is welcomed, even if on a small scale);
2   pursue an ever-more technological approach (controls with automation and IT feedback, simulation, virtual reality, colour therapy rooms, sensory stimulation scenarios and personal implants programmed to give the impression of birdsong and spring sunshine!).

This second scenario has been visited by many science fiction writers, one suspects cynically, rather than enthusiastically. If successful, it would give the 'advantage' of being able to completely disengage from nature. There would be no environmental limit to the height and depth of buildings, and the density of their occupation.

Rendering this question of natural versus synthetic especially relevant is the fact that the majority of global energy use is employed in reducing the impact of the natural environment on us. This suggests that the answer to the question could make a major contribution to sustainability, as well as to the health and well-being of people.

As is customary at the end of scientific papers, we say that there is need for more research. Few would deny that the human response does not feature strongly in the field environmental engineering. Whilst architects may claim that they look after the human interests, at best this has been at a sincere but pragmatic level, and at worst has been the application of personal dogma and patronising principles. It is hoped that this paper and others in this volume will help make the case for a new field of cross-disciplinary study, bringing physics, biology, psychology and sociology into the architecture and engineering of the built environment.

References

Baker, N. and Standeven, M. (1994) 'Comfort criteria for passively cooled buildings: A PASCOOL task', *Renewable Energy*, 5(5): 977–84.

Baker, N. and Standeven, M. (1996) 'Thermal comfort for free-running buildings', *Energy and Buildings*, 23: 175–82.

Brager, G. and de Dear, R. (2000) 'A standard for natural ventilation', *ASHREA Journal*, 42(10): 21–7.

Cawthorne, D. (1994) 'Daylighting and occupant health in buildings', unpublished PhD thesis, University of Cambridge.

de Dear, R. J. (1998) 'A global database of thermal comfort field experiments', *ASHRAE Transactions*, 104(1b): 1141–52.

Edwards, B. (2003) *Green Buildings Pay*, 2nd edn, London: Spon Press.

Fanger, P. O. (1970) *Thermal comfort*, New York: McGraw-Hill.

Fjeld, T., Veiersted, B., Sandvik, L., Risse, G. and Levy, F. (1998) 'The effect of indoor foliage plants on the health and discomfort symptoms among office workers', *Indoor and Built Environment*, 7(4): 204–9.

Guedes, M. (2000) 'Thermal comfort and passive cooling in southern European offices', unpublished PhD thesis, University of Cambridge.

Heschong, L. (1979) *Thermal Delight in Architecture*, Cambridge Mass: MIT Press.

Humphreys, M. A. (1978) *Outdoor Temperatures and Comfort Indoors*, Building Research Establishment, Current Paper 53/78, Watford: BRE.

Humphreys, M. A. and Nicol, J. F. (1970) 'An investigation into the thermal comfort of office workers', *JIHVE*, 38: 181–9.

Keep, P., James, J. and Inman, M. (1980) 'Windows in the intensive therapy unit', *Anaesthesia*, 35: 257–62.

Nikolopoulou, M. H. (1998) 'Comfort in Outdoor Spaces', unpublished PhD thesis, University of Cambridge.

Papairi, K. (1999) 'Daylighting in architecture: Quality and user preference', unpublished PhD thesis, University of Cambridge.

Saxon Snell, H. (1883) *Hospital construction and management*, London: J. and A. Churchill.

Schiller, G. E. (1990) 'A comparison of measured and predicted comfort in office buildings', *ASHRAE Transactions*, 96(1).

Ulrich, R. S. (1984) 'View through a window may influence recovery from surgery', *Science*, 224: 420–1.

Winston, R. (2003) *Human instinct: How our primal impulses shape our modern lives*, London: Bantam.

Wolverton, B. C., McDonald, R. C. and Mesick, H. H. (1985) 'Foliage plants for indoor removal of primary combustion gases carbon monoxide and nitrogen dioxide', *Journal of the Mississippi Academy of Sciences*, 30: 1–8.

Chapter 5

Designing diverse lifetimes for evolving buildings

John E. Fernandez

Materials flows, construction and projecting futures

Introduction

Imagine a building that gracefully accepts and even facilitates positive change of many kinds. Imagine a building that, after construction, continues to be appropriate in use, not only in the immediate present but equally in the distant future — whatever the future may hold. Can a building be designed that leaves its site in a better state of environmental and social balance than before it was built? If such a structure were possible, would it also require radical changes in the way designers and engineers conceive of buildings? How would one go about designing such a building?

Clearly, buildings are not now able to so easily contend with the uncertainties of the future. Most industrialised societies readily accept that the normal lifetime sequence of a building leads to demolition and the landfill. Construction and demolition waste (C&DW) comprises 20–30% of all landfill wastes in industrialised countries. In the USA, for example, 24% of the municipal solid waste stream (MSWS), by weight, is generated by construction (Craven 1994).

Even when buildings are designed to adapt quickly to change, rarely do they succeed. The Centre Pompidou aspired to significant capabilities to alter its internal spaces through movable floors and partitions, but never fully realised the potential (Figure 5.1). As Piano and Rogers wrote and submitted with the competition entry:

5.1
Centre Pompidou, Paris, France

the Centre's internal flexibility should be as large (i.e. great) as possible. In a living and complex organism such as the Centre, the evolution of needs is to be especially taken into account.

(Silver 1994: 24)

The Centre has proven to be an extremely successful building on its merits as a good piece of urban design. The Millennium Dome in London is probably the most extreme example of a building for the future but burdened with such an optimistic projection of a singular future that it now resides without any real function at all — possibly destined for scrap recycling and the landfill (Figure 5.2).

Minimising waste, reducing consumption, recycling materials and other strategies have all been part of a suite of techniques used to improve the efficiency with which the buildings use materials as they are asked to respond to change. This chapter presents another way in which to address the responsible use of renewable and non-renewable materials in architecture through the design of building lifetimes. Intelligent obsolescence through diversified lifetimes will be described in the second half of this chapter.

5.2
Millennium Dome, London, England

Design and change

Designers have always contended with the flux of change. In this age of furious technological, constant political and surprising social change, one should be asking why designers do not attempt to design buildings that can react well to change, even dramatic change of many kinds (Figure 5.3). The design of a building requires that one have a notion of the activities that will take place within the structure and how long the structure should serve to

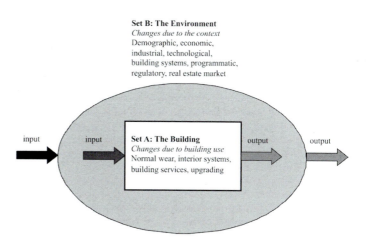

Set B: The Environment
Changes due to the context
Demographic, economic,
industrial, technological,
building systems, programmatic,
regulatory, real estate market

input   input

Set A: The Building
Changes due to building use
Normal wear, interior systems,
building services, upgrading

output   output

5.3

**The context of a changing building
within a changing environment**

shelter those activities. In fact, the most essential condition for architecture is the sheltering of activities over time. This condition remains paramount in the set of expectations for any construction project — what activities need to be protected, and for how long?

In order to satisfy this condition — the sheltering of activities over time — it is necessary to correctly identify the resident activities and designate an appropriate lifetime for the building. If the building does not properly support the activities within it for the prescribed amount of time, the needs of users will not be met. And yet, buildings often belie attempts to comprehensively specify the exact activities within, especially after occupation and into the future. As a result, there may be constant change to the physical systems of buildings. Many buildings bear out the fact that initial construction costs for a new building constitute only one-third of the total cost over its entire life — a significant portion of the remaining two-thirds is due to the costs associated with renovation, repair and rearrangement of interior spaces and systems. Also, projections for an optimal lifetime are often incorrect. As architects often find, many buildings are never used in the ways they intended.

What happens when the use and the appropriate lifetime for which the building was designed changes? Change, in its many forms, will alter the use of a building and impose durability stresses and other lifetime burdens on a building the moment it has been completed for occupation. Three types of changes in buildings are identified here (Table 5.1). They are changes in the function of the space, changes in the load carried by the systems of the building, and changes in the flux of people and forces from the environment (Slaughter 2001: 210).

With the imprecision in predicting use and lifetimes for buildings, a resource problem emerges. The built world consumes enormous amounts of material. Imprecision, or simply misprojection, may cost enormously in material resources.

Table 5.1 **Factors of change**

Type of change	Change factor	Nature of flux
I. Function	1. Programmatic	Changes in the use patterns, spatial arrangements and performance preferences of users. Also fundamental changes in the activity itself.
	2. Process	Changes in the arrangement of space to serve the function due to better logistics planning, process efficiencies or external influences, such as corporate policy.
	3. Ownership	Changes in the ownership entity and general type of user.
	4. Philosophy	Changes in philosophies regarding the use of space; for example, new modes of work and workplace design.
II. Loads	5. Building systems	Changes in acceptable building systems and the loads that are carried. Also includes upgrades to systems to satisfy more stringent performance requirements.
	6. Repair and replacement	Changes in the system used due to repair and replacement needs. New delivery and control devices introduced as part of redesign of building systems.
	7. Systems innovation	Changes in the types and intensity of use of building systems due to innovations in the way in which services are provided for; for example, the use of natural ventilation for summer cooling.
III. Flux of environmental forces	8. Economic	Changes in national, regional and municipal economic base in which facility is located. Also changes in the nature of the business of the activity.
	9. Industrial	Changes in particular industry itself and the composition of groups of interdependent industries serving a particular need.
	10. Technological	Changes in information, material, processing and other technologies surrounding the resident activity.
	11. Demographic	Changes in numbers, composition, ages, education and many other issues pertaining to the population residing in or using the building in some way.
	12. Regulatory	Changes in the structure and emphasis of zoning, building, systems, materials and other relevant codes.
	13. Real estate market	Changes in the property investment and value market. Also, changes in the types of buildings that are best suited to serve particular markets; for example, the smart office building.

Source: Slaughter 2001: 210

Material consumption

The richest countries, with 20% of the world's population, consume 86% of the world's resources. Currently, the USA is one of the largest consumers with 5% of the global population and 7% of its land area while it consumes almost one-third of the world's non-energy materials (Geiser 2001). During the twentieth century, annual material consumption increased by a multiple of almost 20 and consumption of half of these materials occurred within the past 25 years. However, it is now clear that the rest of the world is catching up.

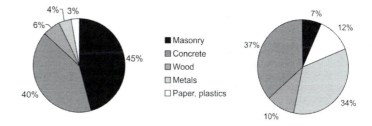

5.4

Average composition of demolition waste in Europe (left) and the USA (right) (Wernick _et al._ 1996 and Kibert _et al._ 2000)

The increase in the rate of materials consumption in the rest of the world is now twice that of the USA (Geiser 2001). Unfortunately, globally 30% of all material consumption is dissipated into the environment — never to be fully recovered. In most countries, construction is the single largest industry in the economy. In the USA, the construction industry employs close to 8 million people and 70% of apparent material consumption (by weight) is dedicated to construction (Wernick _et al._ 1996). C&DW in the USA total approximately 40 million tons per year (Figure 5.4). Only 10% of construction materials consumption is currently recycled. The human species is now consuming natural capital of all forms beyond the carrying capacity of the earth. It has been estimated that 120% of the earth's land area is now necessary to satisfy the demands of the global population (Wackernagel _et al._ 2002).

The construction industry is slowly coming round to understanding the magnitude of its contribution to the depletion of non-renewable resources, the generation of pollutants and the spoiling of land (Moavenzadeh 1994). Proposing better ways in which buildings can change over time will offer the possibility that buildings become much more efficient users of materials, renewable and non-renewable. These more efficient buildings will not only save material and energy that is directly used in their physical systems but also the materials and energy necessary to bring these systems together on the building site, namely material overburdens. These overburdens can amount to many thousands of times the amount of material being delivered to its final use (von Weizsacker 1998: 45–54).

In addition, the industrial context is rife with discarded buildings and misused land and water resources. The ultimate spoiling of the future is toxic land left over from large-scale industrial processes. Research has been conducted that addresses the potential for land reclamation and improvement as a way in which to provide a resource for the future (Ball 1999; Schmidheiney 1992: 374).

In addition, it has been found that buildings are increasingly fulfilling their lifetime requirements within 15–20 years, a dramatic departure from the notion that buildings should be built to last. The Building Owners and Managers Association (BOMA) of Australia estimates that, more and more, various building types are now only needed for 15–20 years (BOMA 2000). The notion of short lifespan buildings increases the need to formulate design strategies and building technologies that promote a responsible turnover.

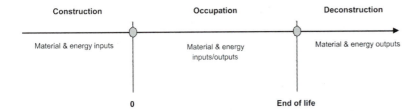

Three primary stages describe the full lifetime of a building. Each stage is characterised by very different materials flows and together they constitute a linear cradle-to-grave sequence (Figure 5.5).

Alternative models include semi and completed closed consumption models. In a schematic sense, there are four modes for materials reuse in the built environment: recycling, reprocessing, reuse and relocation (Figure 5.6) (Crowther 2000: 21).

The ideal condition is one in which the building is part of a closed loop. In a closed loop waste does not exist. Any product from one action will be incorporated into another process. Responsible management of material flows includes:

- reducing material intensity;
- reducing energy intensity;
- enhancing material recyclability;
- reducing dispersion of toxic substances;
- maximising sustainable use of renewable resources;
- extending product durability (NMAB-501 2001).

Governing these strategies is the general interest in reducing, reusing and recycling materials and components from buildings. However, the

5.5
Building cradle-to-grave

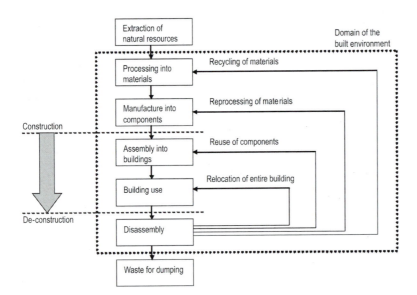

5.6
Materials cycles in construction

mantra of 'reduce, reuse and recycle' has come under criticism for being too limiting in vision and impact (McDonough and Braungart 2002). Therefore, instead of the simple minimisation of resources use in a linear cradle-to-grave model, buildings may be considered to be part of a larger circle of relations; that of material flows and cycles. Buildings are best considered to be part of a larger ecology; an ecology of the built landscape. However, it is necessary to consider this ecosystem within time frames that take into account the longer lifetimes of buildings; 20, 50, 100 years and beyond. Considered in this way, new design strategies can be formulated to assist the construction of buildings that easily allow the flow of materials into and out of them for the purpose of serving their occupants — and society generally — more responsibly. The critical link is made when forging a connection between the useful lifetime of a building and the strategy of closing materials cycles, both technical and biological. Before we can address this linkage, service lives and the projection of lifetimes will be considered.

Design and service lives

Service life, for any manufactured artifact, is the period during which a properly maintained component or system performs at or above predetermined minimum acceptable values. The design life is the expected service life that the properly maintained component or system is meant to achieve under prescribed performance conditions (Nireki 1996). Service life design is the process through which a designer or engineer configures and specifies the components of a system such that the design life may be achieved. There are ongoing efforts to improve the ability of designers and engineers to precisely determine the appropriate design life of buildings[1] (Soronis 1996).

 With the design of a building, as with any physical artifact, the designer determines the location of the material within a technical or biological materials flow cycle, depending on a service life prediction and design life need. If a material has been recycled, it is located further through its functional cycle than a material recently recovered from natural sources. The choice of where to place the material within a cycle is affected by the anticipated service life of the artifact being designed. However, there is always risk in the designation of a particular lifetime for a material, component, system or building. The fact that a building is an extremely complex assembly of thousands of components with varying performance requirements and distinct sets of stressors, means that designating a particular lifetime is more an act of faith than the result of a rational method.

 One method for diversifying this risk is to design a building that allows for a response to an extremely wide range of future conditions. Suppose that a building could accommodate not just three or four likely outcomes, but any possible future. A simple parametric diagram, see Figure 5.7, which shows the two extremes for the future of the building, can define the range of possible physical changes to the building. On the one extreme,

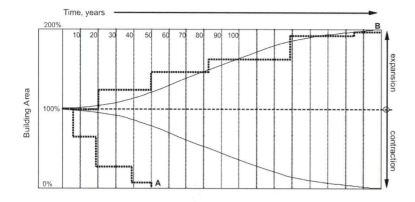

5.7
Parametric building evolution diagram

the building reaches a point in which it is no longer useful to the current or any future user. This condition makes the building completely redundant and results in a loss of space. The other extreme is one in which the building has found continued use and additional further uses. The two conditions are traced through two hypothetical future histories. The former condition calls for an architecture that can be removed from the site (A). The latter calls for an architecture that is durable and easily expanded (B).

At first it may seem that these conditions are mutually exclusive of any single design approach or proposal. Yet, the opportunity lies in the development of an architecture that promotes its own reassessment during its lifetime and thus allows for each future to be possible. Before describing the method of diversified lifetimes, a few words on the difficulties of projection follow.

The fallacy of projection and scenario-buffered design

Projecting future uses through programming is an accepted method for the determination of the optimal use of a new building's space. Programming is, typically, the first step in proposing a design solution for a spatial need. The most common steps taken during a programming phase are:

1   correctly identify the relevant user groups;
2   correctly identify primary personnel within each user group;
3   interview user groups and accomplish additional data collection;
4   transpose data into spatial diagrams, quantitative information and qualitative remarks;
5   write the design brief that will be used by the design team in the design of the new building.

The fundamental problem with programming resides in the fact that it is a projection of the future and nothing less. In so many ways, this future vision is bound to be incorrect. One response to the inability of a typical programming process to capture the correct evolutionary path of the building over time is the process of 'scenario-buffered design'.

Scenario-shift is the type of fundamental change that renders entire buildings obsolete or without reasonable reuse potential. The idea of scenario-shift closely reflects the elements of change listed in Table 5.1. Stewart Brand has described a planning and design process that takes into account the likelihood of future shifts in scenario as 'scenario-buffered design' (Brand 1994). He considers scenario-buffered design a superior method for hedging the future and producing a more flexible, responsive and ultimately responsible building.

> The great vice of programming is that it over-responds to the immediate needs of the immediate users, leaving future users out of the picture, making the building all too optimal to the present and maladaptive for the future. An old saw of biology decrees, 'The more adaptive an organism to present conditions, the less adaptable it can be to unknown future conditions' ... The iron rule of planning is: whatever a client or an architect says will happen with a building, won't.
>
> (Brand 1994: 13)

In contrast to the process of envisioning the correct future, scenario-buffered design attempts to develop strategies by which the building may 'attend to the future'. In addition, Brand also refers to the notion of 'future preservation' advocated by Kevin Lynch (Lynch 1972: 115). Future preservation advocates treating the design of buildings as a service in time. Design is focused on achieving an appropriate durability as well as engineering into the building itself the ability to adapt for any number of future scenarios.

Scenario-buffered design also places value on construction modes that allow for an inconclusive finish to the building. In some instances the best strategy may be a range of conclusions at the completion of the building. In other words, as opposed to completing the building all to the same level of finish, the designers may purposefully leave certain areas 'unfinished', or simply finished to a significantly lesser degree. This would allow the users to determine the appropriate materials to be used, as well as allowing the organisation to monitor the evolution of use of the space and then plan accordingly. Scenario-buffered design is one of the few alternatives to the common practice of programming that directly addresses the high likelihood of an unforeseen future.

Another mode of improving the methods that designers use to accommodate change into their design is the idea that a flexibility scenario should be built into the design brief for building requirements (Prins *et al.* 1993). Designing a flexibility scenario may be an interesting requirement; however, the quality of the solution is highly dependent on the creativity of the designer. In any case, scenario-buffered design and other targeted programming methods still depend on a projection of several discrete 'likely' outcomes. While these are interesting methods, the building industry needs a strategy by which buildings may accommodate change through physical reconfiguration of any kind. This need is addressed by the method of diversified lifetimes.

Lifetime diversity as a method for design

Diversified lifetimes (DL) was developed at MIT as a more comprehensive method for the design of buildings that allow for the future, any future, in a responsible way (Fernandez 2002). DL proposes that any future may be served through the continual re-evaluation of the form and configuration of building systems and large-scale physical assemblies. The reconfiguration may respond to a host of forces that will act on the building through time. The key to the strategy for reconfiguration is not attempting to predict what kinds of forces may act on the building and when they may appear, but providing mechanisms by which the building may respond to any combination of forces, at any time. As a result, DL also does not require that the designers have a predetermined path of evolution for the building, only that the design accommodates evolution. Over the lifetime of the building, both the final form and the activities within may be changed dramatically.

The fundamental directive of a DL strategy proposes that the designer of any architectural project of significant size may reduce risk by distributing a range of building lifetimes throughout the project. This approach is in stark contrast to the typical designation of a single project lifetime for a building. Figure 5.8 illustrates this principle using a simple graph of building areas in percentage of the total (vertical axis) and time (horizontal axis). The rectangle on the left is a building in which 100% of its space has been designated a 50-year lifespan. To the right of the arrow the building has been apportioned, by distinct areas, varying lifetimes. The purpose of such an approach is the reduction of the risk that portions of the built facility will be used inefficiently as a direct result of inflexibility. Incompatibilities between the architectural form and the activities within are eliminated through an evolutionary process that facilitates the reconfiguration of the form, both internal and volumetric. Through a careful distribution of a range of building lifetimes, the facility will engender a more flexible and diverse architecture that evolves with the varying types and levels of stressors present. Figure 5.9 illustrates several representative profiles of different building evolution paths.

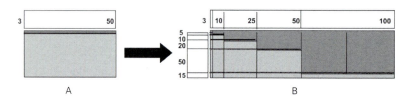

A   B

5.8
Homogeneous lifetime (left) and heterogeneous lifetime diagram (right)

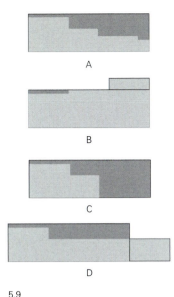

5.9
Various lifetime modes:
A — contraction; B — expansion;
C — disappearance; D — long-life
contraction

DL depends on existing technologies to facilitate the kinds of changes necessary for flexible buildings. Design for disassembly, separation technologies, materials reclamation and recycling, loose-fit detailing, lightly-treading foundations and other technologies, will all contribute to a suite of technologies necessary for building volumes to change over time. DL brings all of these technologies into an integrated process of renewal and change.

In particular, disassembly technologies primarily focused on recovering materials for reuse and recycling play an integral role in the design of the building. While there have been numerous disassembly initiatives globally, the notion of designing into a building its own disassembly specifications has not been fully researched (Kibert and Chini 2000). Closely related is the field of separation technologies (NMAB-487-3 1998). This is another area in which architectural design can advance toward flexible buildings.

Also, general design strategies for increased building flexibility have been formulated, at least in a schematic form. One researcher has organised these strategies into ten clusters of design strategies (Slaughter 2001: 214). They are:

1   reduce intersystem interactions;
2   reduce intrasystem interactions;
3   use interchangeable system components;
4   increase layout predictability;
5   improve physical access;
6   dedicate specific area/volume for system zone;
7   enhance system access proximity;
8   improve flow;
9   phase system installation;
10   simplify partial/phased demolition.

DL intends to more easily allow a building to adapt over time through changes to:

- the overall volume of the building;
- the footprint of the building;
- the arrangement of interior spaces and physical elements of the architecture;
- the overall architectural typology of the building;
- the diversity of spaces and materials used for the building;
- the overall commitment of capital investment in the facility;
- the composition of the remaining lifetime profile of the facility;
- the overall durability of the facility; and
- the capacity of the building to accept new programmatic uses.

Case study: British Petroleum Headquarters, Aberdeen Scotland

The principles outlined above are applied to the design of a new Exploration Headquarters building for British Petroleum in Dyce, just outside Aberdeen, Scotland.[2] The building is intended to house the various business units that are responsible for operations in the North Sea region. The new building will replace the existing headquarters, now deemed unusable by the current management, and provide 16,500 m² of office space (British Petroleum 2001). A large leisure building, housing a skating rink and other recreational activities, currently occupies the site along with a moderate amount of surface parking.

Scotland is an interesting context in which to develop an experiment in sustainable development using diversified lifetimes. The Scottish Parliament placed sustainable development high in their listing of priorities as early as February 1999. The Scottish National Waste Strategy articulates a programme for resource use, adopted in December 1999. The coordination of these efforts is the responsibility of the Ministerial Group on Sustainable Scotland, comprised of six ministers and two external members (DETR 2001). The current UK target for industrial and commercial waste reduction is landfill flows at 85% of the 1998 levels by 2005. While this may seem conservative, it will take significant changes in typical construction practice to achieve.

The construction industry is a very large part of the Scottish economy and accounts for a great deal of the environmental impact of industry. The gross output of construction is valued at £5.6 billion per annum, 6.5% of Scotland's gross domestic product. Almost half of this activity is dedicated to the repair and maintenance of existing buildings. In the UK as a whole, 40–50% of carbon emissions can be attributed to the activities of construction (Sustainable Development Team 1999).

The project is an ideal study for the projection of various evolutionary modes during the building's lifetime. As a result of uncertainty in the future of the business of energy and fossil fuels, BP has predicted that their production in this region will decrease dramatically within the next 15 years. The company has articulated the possibility that BP will leave the area altogether once production slows to a trickle. Then again, predictions are a risky business. In any case, given the context for the future, the building will most certainly need to respond to rather dramatic changes within a very short time.

However, this case study only has sufficient space in this chapter to demonstrate one mode of change — that of contraction. Given BP's predictions, the overall state of the regional economy, and the high vacancy rate for commercial office buildings, economic contraction may be the future reality of Dyce and Aberdeen.

The site for the new building is located 300 m southwest of the complex of buildings that comprise the current headquarters. The buildings are located in the heart of the industrial sector of Dyce. The regional airport is within 0.5 miles and Aberdeen is 8 miles to the southeast.

A diversified lifetime strategy is applied in the following steps:

1   Spatial needs assessed: programming, scenario-buffered interviewing and calculation of lifetime spread for the building.
2   Lifetime projections assembled: discrete lifetimes defined and assigned to specific areas of the building.
3   Correlated and uncorrelated risks listed: both types of risk are listed and weighed against the widest possible paths for the evolution of the building over time.
4   Market issues addressed: end conditions are assessed in terms of the value needed for an economic exit strategy.

Placing the projected lifetime use of the building at the centre of the bell curve shown in Figure 5.10, a distribution of lifetimes ranging from 1 year to about 45 years can be considered.

Figure 5.10 shows a bell curve distribution for a typical building of 50 years' life, and alongside it (as a dashed line), a building of short duration — 25 years. Each has a distinct spread of lifetimes. Given this lifetime distribution, a table can be formulated that specifies lifetimes for discrete portions of the building (Table 5.2).

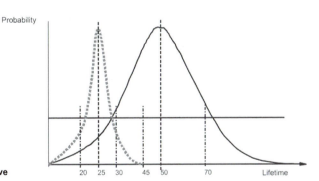

5.10
Lifetime bell curve

Table 5.2 **Lifetime distributions**

Building	Area: %	Area: ft^2	Lifetime: years
Building 1	16	28,800	3
Building 2	10	18,000	10
Building 3	8	14,400	20
Building 4	16	28,800	25
Building 5	12	21,600	25
Building 6	12	21,600	30
Building 7	6	10,800	35
Building 8	20	36,000	45
Totals	100	180,000	

EVOLUTION OVER TIME

Applying these percentage lifetimes to the proposed schemes being considered for the site yields the following results: the scenario of a contraction of the building down to a core of 20% of the original built area. Figure 5.11 illustrates the contraction of the building to its ultimate lifespan of 45 years. For a commercial building this is a very short lifespan. However, for a building within the difficult economic and industrial context of the Aberdeen/Dyce region, the ability to responsibly contract may equally well serve both BP and the community. As it stands, the existing BP Headquarter Building does not have this ability to change and will need to be demolished or adaptively reused; both processes that are highly energy intensive.

As described above, the project is dependent on a range of technologies, from disassembly and separation detailing to materials reclamation and recycling. The design described here is organised with two primary principles in mind:

1   the building itself forms the armature through which materials and assemblies of varying lifetimes may be moved for reconfiguration, reclamation and recycling (see Figure 5.14);
2   the building may change in a number of ways depending on the space needs of the original client and subsequent stakeholders (see Figures 5.15 and 5.16).

Future stakeholders include additional owners of the facility, co-owners, rental tenants and other parties that may come to occupy and use the building. Through these alternative uses, the building may be transformed to most readily suit the current users. Afterwards, the building form may again be reconsidered. A building that is able to disassemble will provide the materials and detail connections that will allow for these kinds of changes. However, a building that is meant to reach obsolescence at varying times will insure that the users are kept aware of the need to reassess space needs and organisational patterns. As individual portions of the building reach their obsolescence, the users may decide to discard these

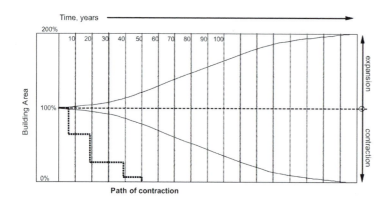

5.11
Lifetime path for diversified building under contraction stresses

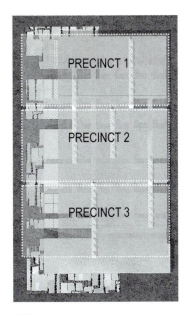

5.12
Precincts

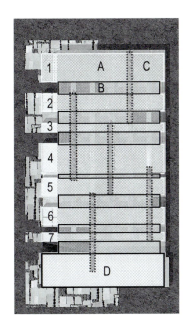

5.13
Function areas

portions through materials reclamation and recycling or they may decide to rebuild the volume, with the ability to update and improve the building's focus on the particular needs of the inhabitants.

A building of this sort also contributes to an overall lowering of the embodied energy in the construction. With shorter lifetime materials, embodied energy decreases. While there may be good reasons to design portions of the building that are designated with lifetimes beyond the stated ultimate lifetime of the building (the area to the right on the bell curve in Figure 5.10), the overall embodied energy of the building has been lowered. As a result, not only is the building more flexible, but it is also embodied with a lower energy level than a building of homogeneous lifetime.

The danger here is avoiding a great deal of construction during the evolutionary process. The energy consumed in transportation of workers and equipment to the site will soon outstrip any embodied energy gains from the lower-durability components of the building. This is the reason why it is important that the building itself should serve as the primary materials handling equipment and that a workshop for materials reclamation and recycling is provided on site.

The building is composed of three precincts of varying lifetimes shown in Figure 5.12. Within each precinct the office space is housed within seven rectangular bars of varying widths (A), the bars are separated by daylit atrium spaces of varying widths (B), and the two are linked by perpendicular light and air shafts (C) that serve to bring natural light into the office spaces and act as air outlets for naturally ventilated cooling (Figure 5.13). Finally, the assemblage of buildings is anchored, on the south end, with a volume (D) that houses public amenities, larger conference rooms and a materials reclamation workshop.

The building facilitates disassembly by enlisting the super-structure as a mechanism for removing and moving materials from their original location within the building back to the on-site materials reclamation workshop. The paths of travel for materials are shown in Figure 5.14. These paths correspond to the necessary structure at the atrium edges and allow many modes of disassembly and reconfiguration. Eventually, materials are brought back to the workshop for reclamation, recycling and distribution to next use.

Two modes of contraction are shown in Figures 5.15 and 5.16. In the first instance, the building retracts from the northern edge over time (at top of image), first from its edges (A) and then from the centre down (B) to the workshop. In Figure 5.16, contraction occurs through the selective disassembly of discrete areas of the building. Both modes, and combinations of each, are possible.

The following sequence of images (Figure 5.17) shows the building as it retracts from its northern edge over time. The images are derived from an animation that shows the building changing over time to respond to a need for smaller overall office space.

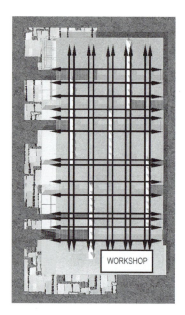

5.14
Disassembly paths

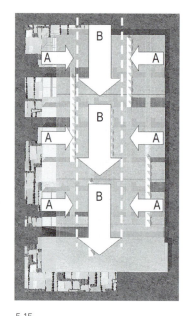

5.15
Contraction mode 1

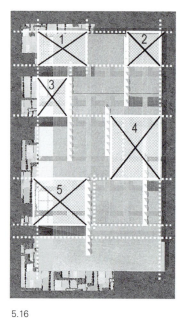

5.16
Contraction mode 2

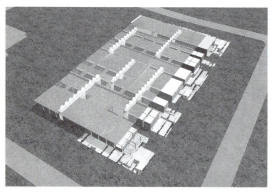

A

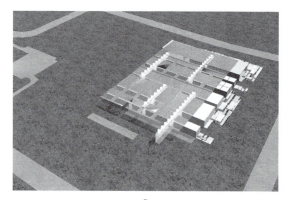

B

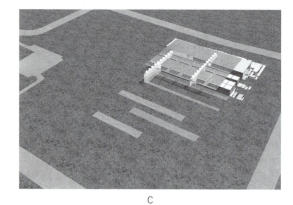

C

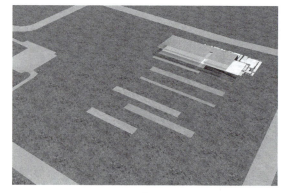

D

5.17
Contraction sequence, mode 1

Conclusions

Disregarding the need for change in buildings necessarily burdens the process of design, construction, occupation and demolition with inefficiencies in material and energy use. It also does not allow for processes of renewal and environmental regeneration. Implementing a strategy for diversifying the lifetimes of buildings and therefore providing a spectrum of good and responsible futures allows for the creative use of resources while enriching the spatial environment with increased diversity. In addition, allowing for change through design also creates value by making the constructed building more agile and better able to respond to unexpected change. This value can be translated into future savings for the owner of the facility and better materials management for the community and society at large.

The process introduced here requires a careful assessment of resources, future trajectories and technological solutions. The process also requires an attention to design for disassembly at both the building and detail levels. A tectonic that is based on change is a very different tectonic than that based on the idea of a monolithic, unchanging building. For industrial, commercial and other types of buildings affected by quick and dramatic changes in the environment, technologies for change should be considered.

In addition, technologies and building systems that allow for change also require a diversity of form and materials. This diversity adds a richness that contributes to the building's potential to become an integral component of the urban landscape.

Notes

1   Most notably the CIB W80/RILEM 140-TSL, a collaborative effort between the CIB (International Council for Building Research and Documentation) and RILEM (International Union of Testing and Research Laboratories for Materials and Structures), the CIB W90, Working Commission on Design for Durability, and Working Group 9, Design Life of Buildings, in Subcommittee 3 of ISO Technical Committee TC59, Building Construction.
2   The project is part of a series of research studies done through the sponsorship of the Cambridge MIT Institute, an organisation that promotes joint research projects addressing issues of the environment. The building was designed by the London firm TP Bennett as part of a feasability study.

References

Ball, R. (1999) 'Developers, regeneration and sustainability issues in the reuse of vacant industrial buildings', *Building Research & Information*, 27(3): 140–8.

Brand, S. (1994) *How Buildings Learn*, New York: Penguin Books.

British Petroleum (2001) *Confidential Design Team Brief for new Exploration HQ Building*, September, internal BP document.

BOMA (2000) *Review of Business Taxation: Comments from the Building and Investment Industry*, Australia: Building Owners and Managers Association.

Craven, D. J., Okralig, H. M. and Eilenberg, I. M. (1994) *Construction waste and a new design methodology*, Sustainable Construction, November 6–9, Tampa, Florida: CIB.

Crowther, P. (2000) 'Chapter 2: Building Deconstruction in Australia', in Task Group 39 *Overview of Deconstruction in Selected Countries*, CIB Report, Publication 252, August, CIB.

DETR (2001) *Transport and the Regions, Achieving a better quality of life, Review of progress toward sustainable development, Government Annual Report 2000*, London: Department of the Environment, Transport and the Regions.

Fernandez, J. (2002) 'Diversified Lifetimes: Orchestrated Obsolescence for Intelligent Change', *Thresholds 24*, Cambridge Mass.: MIT.

Geiser, K. (2001) *Materials Matter, Toward a Sustainable Materials Policy*, Cambridge, Mass.: MIT Press.

Kibert, C. and Chini, A. (2000) *Overview of Deconstruction in Selected Countries*, CIB Report, Task Group 39, August: CIB.

Lynch, K. (1972) *What Time Is This Place?* Cambridge, Mass.: MIT Press.

Moavenzadeh, F. (1994) *Global Construction and the Environment*, New York: Wiley Interscience Publication.

McDonough, W. and Braungart, M. (2002) *Cradle to Cradle, Remaking the Way We Make Things*, New York: North Point Press, Farrar, Strauss & Giroux.

Nireki, T. (1996) 'Service life design', *Construction and Building Materials*, 10(10): 403–6.

NMAB-501 (2001) *Materials in the New Millennium*, Proceedings of the 2000 National Materials Advisory Board Forum, Washington, DC: National Academy Press.

NMAB-487-3 (1998) *Separation Technologies for the Industries of the Future*, Panel on Separation Technology for Industrial Reuse and Recycling Committee on Industrial Technology Assessments, Washington, DC: National Academy Press.

Prins, M., Bax, M. F. T., Carp, J. C. and Templemans Plat, H. (1993) 'A Design Decision Support System for Building Flexibility and Costs', in H. Timmermans (ed.) *Design and decision support systems in architecture*, Netherlands: Kluwer.

Schmidheiney, S. (1992) *Changing Course*, Cambridge, Mass.: MIT Press.

Silver, N. (1994) *The Making of Beaubourg, A Building Biography of the Centre Pompidou, Paris*, Cambridge, Mass.: MIT Press.

Slaughter, E. S. (2001) 'Design strategies to increase building flexibility', *Building Research & Information*, 29(3): 208–17.

Soronis, G. (1996) 'Standards for design life of buildings: utilization in the design process', *Construction and Building Materials*, 10(7): 487–90.

Sustainable Development Team (1999) *Down to Earth: a Scottish perspective on sustainable development*, Edinburgh: The Scottish Office. Online. Available www.sustainable.scotland.gov.uk (accessed 18 October 2003).

von Weizsacker, E.U. (1998) 'Dematerialization — Why and How?', in P. Vellinga, F. Berkout, and J. Gupta (eds), *Managing a Materials World*, Netherlands: Kluwer.

Wackernagel, M., Schulz, N. B, Deumling, D., Linares, A. C., Jenkins, M., Kapos, V., Monfreda, C., Loh, J., Myers, N., Norgaard, R. and Randers, J. (2002) 'Tracking the ecological overshoot of the human economy', *Proc. of the National Academy of Sciences*, USA, 99(14): 9266–71.

Wernick, I. K., Herman, R., Govind, S. and Ausubel, J. H. (1996) 'Materialization and Dematerialization: Measures and Trends', *Daedalus, Journal of the American Academy of Arts and Sciences*, 125(3): 171–98. Table 5.1 Factors of change (source: Slaughter 2001: 210).

PART 3

Urban

Chapter 6

Urban diversity

Koen Steemers, Marylis Ramos and Maria Sinou

Introduction

The urban context provides a rich and varied environment that influences the way we use urban spaces (movement, sequence, activity) and how we feel in them (stimulation, visual/thermal/aural comfort). This chapter introduces the interrelationships between urban form, microclimate and comfort. It draws on recent research of monitoring, surveying and modelling urban thermal characteristics and proposes a method of mapping urban diversity.

In environmental terms our cities are frequently described negatively and associated with pollution, poor health, stress and overcrowding. In practice, many cities have developed in a piecemeal and evolutionary manner, planned and designed over many centuries and involving countless individuals. Although the result is that planning has typically been incremental and ineffective to adapt to new pressures, such as, for example, transport, the result is a richness and variety of urban environments — both in positive and negative terms. These qualities affect our perception of urban spaces and how we use them. The purpose of this chapter is to explore the nature of this environmental diversity with a particular focus on the thermal conditions, by defining the links between urban form and microclimate, and to demonstrate how our perception is influenced by it.

The first part of this chapter addresses the urban microclimate with a particular reference to its diversity rather than the absolute conditions. This is achieved primarily by the study of monitored data. The second part reveals how users of outdoor spaces respond to the urban microclimate, as exposed through a combination of surveys and environmental monitoring. Finally, the chapter concludes by proposing a mapping of environmental diversity, showing how this can be used to understand and inform urban design.

The urban microclimate

It is well understood that the urban climate varies significantly from the synoptic climate, and that this is caused by a combination of anthropogenic and physical factors. The reduction of vegetation, the increase of thermally massive and non-porous surfaces, the presence of drainage systems and the emission of heat and pollution all contribute to altering the urban microclimate. Perhaps most interesting, at least from a planning and design perspective, is that the geometry of urban form also has an important effect. Geometric factors include density, surface-to-volume and height-to-width ratios of urban buildings and spaces. What is perhaps surprising is to find that such complex sets of interactions, between form and microclimate, are typically reduced to a discussion of temperature difference — by way of the 'urban heat island effect' — or pollution levels, etc. Such metrics are expressed as the difference between the rural environs and the urban areas, typically measured under conditions where such differences are greatest. Thus the heat island effect will typically be measured on still and clear summer evenings when maximum values can be recorded.

There are three issues related to such analysis of the urban microclimate:

1   it gives the impression that the urban environment is on average different without revealing the dynamic range;
2   the relative significance of this overall difference compared to smaller scale variations is not clear; and
3   the effects are almost always discussed as negative terms.

In technical terms, such work tends to focus on the 'urban boundary layer' which can be defined as being immediately above the rooftops of buildings and affected by the nature of the urban 'surface'. This is of particular interest to climatologists. The 'urban canopy layer' relates to the spaces in the streets and between buildings, and is of particular interest in terms of conditions experienced by people and buildings. The distinction between the two is critical: the boundary layer is typically well mixed (Oke 1987: 300) whereas the canopy layer 'is characterised by considerable complexity, mainly deriving from the convoluted nature of the active surface' (Oke 1987: 284). It is this complexity that we are interested in here.

In his seminal book *The Climate Near the Ground*, Geiger (1966) refers to variety on the same page but in apparently contradictory terms:

> Where man begins to achieve dominance, the infinite variety of undisturbed creation is disturbed.

And yet:

Men are forever creating new kinds of microclimates. Every building constructed displaces the original climate of its site, creating a warm, sunny and dry climate with a southern exposure on the one hand, and a shady, cool and damp northern climate on the other.

(Geiger 1966: 480)

Geiger goes on to state that 'Taken over the year as a whole, a city is warmer than the surrounding countryside, the difference being in the order of 1 deg.' (Geiger 1966: 492). More recent studies refer to maximum values of as high as 12–14°C (Oke 1987 and Santamouris 2001). How are we as planners and designer to use this kind of data? Further unravelling is needed to understand the implications with respect to urban form and perception.

The measurements and predictions of the urban heat island are typically carried out to reveal peak differences, in order to demonstrate the physical forces and extent of the effect. Oke refers to the 'maximum heat island intensity' in his correlations between the heat island and urban geometry, as follows (Oke 1987: 293):

$$\Delta T_{(max)} = 7.54 + 3.97 \ln(H/W) \tag{6.1}$$

Oke's assessment shows that the narrower a street canyon, or the larger the height to width ratio (H/W), the greater the maximum heat island intensity ($\Delta T_{(max)}$). He concludes that 'the urban geometry is a fundamental control on the urban heat island' (Oke 1987: 293). On the same page he goes on to state that 'Humans however find the added warmth to be stressful if the city is located in an already warm climate'. The conclusion would seem to be that narrow streets are the wrong approach for hot climates as this would increase already high air temperatures. And yet we know that traditional hot-arid cities are very compact with deep streets and courts (Figure 6.1). The reason for this apparent contradiction is that the use of maximum heat island intensity is overly simplistic, and that it is in fact the dynamic diurnal and spatial effects that are critical.

The maximum heat island intensity typically occurs after sunset — at about 9 p.m. Yet there is also the potential for a 'negative heat island' at the hottest time of the day at about 3 p.m. At this critical point in the day it is possible for the urban temperature at the bottom of the urban canopy layer to be cooler than the rural temperatures, providing much needed protection in a hot climate, despite the fact that the urban boundary layer will be hotter. By the time the heat island intensity (i.e. the increment above ambient) is at a maximum, the ambient temperatures will be lower and thus less critical in terms of comfort or stress. This argument clearly suggests that the variations, at least in temporal terms, are critical, offering the potential for cooler conditions at the hottest time of day in return for a temperature increment when air temperatures are cooling down.

6.1
The deep streets and courts of, for example, Granada in southern Spain seem to contradict the logic of urban heat island effect where narrow streets create higher urban temperatures

If we again take the scenario of the hot-arid climate, then a simplistic appreciation of the heat island intensity might lead us to assume that it is better to provide wide, open streets and spaces to minimise thermal stress. However, in the context of human thermal comfort, it can be argued that the presence of shade — as well as shaded masonry surfaces — will be significant. The opportunity to walk out of intense sunlight may offer far greater benefits to comfort than the disadvantage of some increase in the average air temperature. In numerical terms, a change of 1°C in air temperature can be compensated for by an inverse change in horizontal radiation of approximately 70 W/m^2 (Szokolay 1980: 274). Considering that 1000 W/m^2 of direct solar radiation would not be uncommon, this equates to a 14°C air temperature increase — well above typical heat island intensities measured (which as we have seen would occur at night and thus with no solar radiation present). Therefore, in this instance, it is safe to say that shade is a far more significant factor than heat island effects. It is also evident that shade is determined by built form and solar geometry. A similar case could be made for cold climates where the heat island benefits of a compact form are offset by a lack of access to solar radiation. This demonstrates that spatial diversity can play a significant role in terms of the urban environment.

Having established that both temporal and spatial factors will influence performance of the urban environment, the next stage is to explore the extent and significance of diversity within the urban fabric.

Form and diversity

The study of temperature variations through cities has been part of the investigations into the urban heat island effect. Such work not only reveals that temperature increases tend to be at a maximum in the urban core and reduce more or less concentrically outwards, but also shows variations in temperature differences. Such variations are particularly pronounced near parks where temperature reductions of approximately 3°C compared to adjacent urban conditions have been measured (Santamouris 2001: 150). This is confirmed with computer analysis, showing not only similar reductions but also subtle variations across a park and adjacent urban context in relation to wind effects (Dimoudi and Nikolopoulou 2003). One might suspect more diversity when one looks even more closely at urban form. Some urban spaces may be exposed to direct sunlight for a large part of the day, whilst others may be shaded or vegetated, so that in the same street a diverse range of thermal conditions might be found.

The key questions are: what is the nature and extent of this diversity, and what is the relationship to built form? The work of Oke (1987) and others has already indicated that the temperature conditions in streets and squares is related to the degree of enclosure, expressed as height-to-width ratios or sky view factors. A similar definition might be used to define the thermal characteristics of semi-enclosed spaces such as colonnades and loggia. Potvin, in Chapter 8 of this book, discusses glazed arcades, prominent in cooler climates; here we discuss intermediate environments which have a solid overhead plane, common both in warm and cold climates (Figure 6.2).

6.2
An example of an intermediate urban space from Bologna, Italy

The selective climatic response of such spaces is evident: colder high latitudes benefit from wind and precipitation protection and solar access, whereas warm southern latitudes need shade from the high summer sun.

For the purpose of this chapter we will look at two urban contexts: Kastro on the Island of Siphnos in Greece and Cambridge in the UK. Kastro has a fairly high density urban grain with street height-to-width (H/W) ratios of between 2.0 and 6.0 (average of about 3.0). Using Oke's correlation (equation (6.1) above) this would result in a maximum heat island intensity of 10.0 to 14.5°C. In fact, Oke's graphic scale of H/W ratios only goes as high as 3.5 (Oke 1987: 292), suggesting that Kastro is much denser than typical urban forms. By comparison, Cambridge has H/W ratios typically ranging from 0.5 to 2.0, averaging at about 0.8. This is more typical for European cities and results in a maximum theoretical temperature increase of 6.5°C.

Looking more closely at the two urban areas one can identify a range of intermediate spaces: loggias, passageways, cloisters and colonnades of different proportions and orientations. Their spatial characteristics, having opaque overhead surfaces, immediately make it clear that a simple H/W ratio is not an appropriate indicator of their forms, but that an alternative measure is required in order to compare form with thermal performance. The measure that we propose is the 'Degree of Enclosure' (DoE). This is defined as the ratio of the sum of all surfaces (closed and open (A_{total})) to the sum of the opening areas (A_{open}), as follows:

$$DoE = \Sigma A_{total} / \Sigma A_{open} \qquad (6.2)$$

Thus, one can define a number of generic intermediate spaces with their degree of enclosure assuming cubic proportions as indicated in Table 6.1.

A series of spaces with different degrees of enclosure were monitored in Kastro during the summer period. Initial results suggest that

Table 6.1 **Generic types of semi-enclosed spaces and their degree of enclosure**

Type of space	Plan diagram	Degree of enclosure
Porch		6.0
Arcade		3.0
Portico		2.0
Undercroft		1.5
Open space		1.0

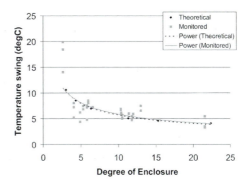

the extent of the diurnal temperature fluctuations correlates inversely with the degree of enclosure. Thus, the less enclosed an intermediate space is, the greater the temperature variations. This makes sense and follows the trends that have been found at the larger scale related to the urban heat island effect.

Recent monitoring of urban spaces in Cambridge, UK by the authors has shown that a temperature range of 6.4°C was measured on a spring morning, with the coolest being in an open park (2.1°C) and the warmest in a cloister (8.5 °C). The ambient temperature at the time was 0.4°C. This suggests that different types and combinations of spaces provide an increase in thermal diversity.

The next step in the analysis is not simply to measure differences but to gain an insight into the physics in order to explain principles of the diversity that has been measured. A model was developed, based on the CIBSE Admittance Method, to explore the temperature implications of intermediate space design characteristics. The temperature swings predicted by the model are compared to the monitored data from Kastro and Cambridge intermediate spaces of various DoE in Figure 6.3. The trend-line drawn through the monitored dataset shows a modest correlation, with an r-squared value of 0.53. More significant is the fact that the monitored trend-line matches the theoretical data produced in the model, resulting in the following equation:

$$T_{swing} = 17.3 \times DoE^{-0.48} \tag{6.3}$$

So, on the one hand there is a fairly wide diversity of thermal conditions found in the monitored spaces, but on the other hand there is a clear trend showing that DoE affects temperatures.

The temperature difference between a space with a small temperature swing and the ambient conditions is likely to be noticeable for more of the time, providing increased diversity. Thus, as ambient conditions get cooler, an intermediate space with a high degree of enclosure will provide a warmer environment. Conversely, as ambient temperatures increase, the space may be expected to remain cooler.

The theoretical curve matches the average monitored curve so well that it provides a good opportunity to explore what causes the deviations from the trend-line. One factor will evidently be the ambient conditions. However, it is more interesting to explore the thermal diversity generated by built form: specifically the volume of the space, its proportions and its orientation.

A change in the volume of a space, whilst maintaining all other variables constant, has an effect on the theoretical temperature swing. Although the volume becomes more influential for small spaces, overall only a modest 2°C alteration can be modelled. Changing the proportions also has even smaller effects, nearly undetectable within a range from 1:1 to 1:5 of the height-to-width proportions of an arcade where the DoE is kept constant. However, altering the orientation of an arcade does indicate that this variable is an important factor. Figure 6.4 below shows that the orientation of an arcade with two opposite open ends influences the temperature swing by 3.5°C, and that a west–east orientation creates the largest range of temperatures. One could imagine that an open-sided, west-facing colonnaded space or portico would provide yet greater variations to the temperature swing.

Even based just on the theoretical studies presented here, one can see that the basic formal characteristics of streets, squares and inter-mediate spaces will result in significantly diverse thermal conditions. This is supported by the monitored data from the case studies. The next question that is raised is how do people respond to the thermal conditions and diversity in an urban environment?

Outdoor comfort

It is, in theory, possible to predict thermal comfort as a function of environmental parameters (such as temperature and air speed) and personal variables (such as clothing level and metabolic rate). However, as comfort has been defined as 'that state of mind which expresses satisfaction with the

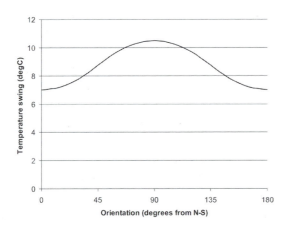

6.4
Predicted variation in temperature swing as a function of orientation for an arcade type of intermediate environment (the orientation is defined as the direction of the axis from one end to the opening at the other end of a linear arcade)

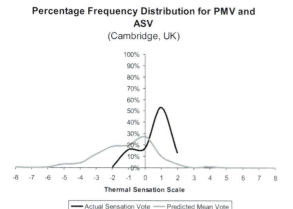

Percentage Frequency Distribution for PMV and ASV
(Cambridge, UK)

Thermal Sensation Scale

— Actual Sensation Vote — Predicted Mean Vote

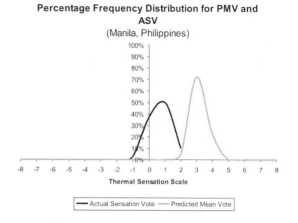

Percentage Frequency Distribution for PMV and ASV
(Manila, Philippines)

Thermal Sensation Scale

— Actual Sensation Vote — Predicted Mean Vote

6.5

Comparisons of the Predicted Main Votes (PMV) and Actual Sensation Votes (ASV) for Cambridge (left) and Manila (right)

thermal environment' (ASHRAE 1992) it is suggested that comfort is related to both physiological and psychological variables. It would thus be arguable that the perception of the users of spaces, as opposed to the physical parameters, is the appropriate measure of comfort. Alternatively, ISO 7730 (1994) defines thermal comfort in purely physiological terms, such as 'thermal neutrality' and the maintenance of the body's energy balance. No extent of environmental measurement will be able to reveal the 'state of mind'. Instead, carefully constructed questionnaires, alongside detailed monitoring, are the vehicle used here to reveal people's experience of the thermal environment. People's perception of comfort, in a range of urban spaces, is subsequently compared with theoretical predictions of physiological comfort with reference to ISO 7730 (1994).

For the purpose of this chapter we will discuss survey results from two urban spaces in Cambridge, UK, and one urban park in Manila, Philippines. The reason for discussing these case studies is that they represent a range of conditions from compact urban square in a temperate climate, to more open space in a hot-humid context.

The key results, comparing the actual with the predicted comfort votes, immediately highlight a discrepancy between the two (Figure 6.5). During warm conditions, particularly noticeable from the Manila surveys, the theoretical figures, based on ISO 7730 calculations using detailed monitored and observed data, suggest that the users of the spaces should be far too hot. Conversely, during cooler summer weather the theory indicates that people should be colder than they actually report. Another way of presenting this information is that the percentage of people expressing dissatisfaction was far lower than one would predict from the environmental data recorded. For Manila, the predicted percentage dissatisfied (PPD) is 100% whereas the actual percentage dissatisfied (APD) is 43%. For Cambridge the figures are 43% and 6% respectively, confirming the discrepancy between the theoretical model and actual comfort perception. Intriguingly, in both Cambridge and Manila the actual comfort vote peaks at +1 (defined as 'slightly warm' or 'comfortably warm') despite the very different climatic conditions. During the summer surveys in Cambridge globe temperature ranges from 17–33°C, compared with 33–39°C in Manila (air temperatures are from 16–30 and 30–33°C respectively).

So we have two issues here:

1   the theory diverges significantly from the responses; and
2   the responses are similar for quite different climatic conditions.

The first point is the most important because we speculate that offering diversity is a more significant parameter than ensuring optimum thermal conditions in physiological terms. This is discussed further below. The second issue may also be influenced by the greater degree of freedom of choice found in urban environments compared to internal environments, on which conventional comfort theory is based. It is, furthermore, likely that acclimatisation is a critical factor for the Cambridge and Manila results being similar. Assuming monthly average temperatures, in England during the summer the neutral temperature, that is the temperature at which most people will feel comfortable — neither too cold nor too warm — is 21°C. In Manila the neutral temperature is in the order of 27°C. This goes some way to explaining why in two quite different climates the comfort responses can be broadly similar; slightly warm. Further explanation for the discrepancies may be attributable to cultural differences, though this remains a speculative point here.

The key question remains: why are the predicted physiological comfort levels so different from the actual responses? Or, putting it another way, why do most people say they are comfortable when technically they should be more often too cold or hot? The assertion here is that increased freedom of choice is a fundamental factor. This assertion correlates directly with the research on 'adaptive opportunity' in the indoor comfort field (see Chapter 4 of this book by Nick Baker). The ability of building occupants to regulate their thermal conditions to suit their preferences will influence the comfort temperature range: more choice results in a wider comfort range. With respect to the interior context, adaptive opportunities range from interactions with the building fabric (opening windows, adjusting blinds, etc.) to personal variables (altering position, adjusting clothing, taking warm or cold drinks, etc.). In the urban environment the 'occupants' have the same, if not a wider range of personal variables. These include the choice to be in a particular place (which is typically not available to an office worker); the ability to move within a space (e.g. to find shade); the metabolic rate (e.g. from lying on the grass to jogging); the potential availability of hot or cold drinks (i.e. a café); greater freedom in clothing (office dress tends to be more constrained), etc.

Most important amongst the urban personal variables is the choice to be in any one particular space. This is confirmed by survey data which

6.6
Il Campo in Siena, Italy offers a unique environment to walk or to sit on the beach-like sloping piazza, in either sun or shade

demonstrates that perceived outdoor comfort is related to the degree of choice people have to be there: those needing to wait (e.g. for a friend or public transport) express twice the amount of dissatisfaction with thermal comfort compared to those with a choice to move. Interestingly, whether such choices are exercised is not critical; simply the knowledge that they exist creates a sense of control and therefore a greater tolerance. Nikolopoulou and Steemers (2003: 98) have suggested that 'perceived control' is the most significant psychological parameter determining outdoor comfort. Clearly, in order for choices to be perceived and potentially exercised there must be a sufficient diversity of environments available (Figure 6.6).

The importance of choice and diversity is supported by psychologists who claim that 'higher organisms engage in an active process of seeking this variability if not found in the immediate surroundings' (Stea 1976: 41) and that in any situational context, the individual attempts to organise his physical environment so that it maximises his freedom of choice (Proshansky *et al.* 1976). Biologists, too, support the notion that the optimum environment should not be constant but should provide an optimum range and frequency of change to be desirable for health and well-being (Potter 1971).

Mapping environmental diversity

So far we have suggested that the urban environment is potentially more diverse in thermal terms, and particularly that intermediate spaces offer an opportunity to extend diversity. We have also discussed how diversity is a particularly important factor in terms of increasing thermal comfort in outdoor spaces. What this section presents is a way of mapping spatial environmental diversity in cities.

As thermal comfort is dependent on air temperature, radiant conditions and air movement, the starting point for plotting the urban environment is to use temperature variations, sunlight and wind as the key parameters. We have already demonstrated that temperature fluctuations can be related to the degree of enclosure of an urban space. In this instance, we will use an equivalent metric: the sky view-factor, which is a relatively simple geometric parameter to determine. In terms of solar radiation it is a simple exercise to map shadows in relation to urban geometry. Finally, the urban wind field presents a much more complex condition that, for the purpose of the exercise, we will simplify using the notion of wind shadows in relation to prevailing wind directions. So now we have, at least in principle, the ability to map temperature, radiant and air movement conditions. For the latter two, solar and wind, we will determine the presence or lack of the variable in an urban space: sun or shade, windy or still. For temperature variations, the gradations are more subtle, dependent on the sky view factor, though it is possible to introduce artificial thresholds or bands in the data for ease of discretisation. Each is discussed below.

Digital elevation models (DEMs) are used to provide a simple representation of three-dimensional urban built form. The density of pixels is used to represent the height of buildings, such that white is the ground or datum level, and black represents the tallest building (Figure 6.7). Using computer software, such as MatLab, it is possible to perform analysis on the urban forms to produce a map of, in this instance, sky view factors (SVF), solar and wind shadows.

Thermal analysis can be based either on the maximum heat island intensity (ΔT_{max}) defined by Oke, or by a comparison of globe temperature swings, as discussed earlier. As Oke's correlation represents extreme conditions, the proposal is that temperature swings are a more useful metric to relate to diversity, particularly as the monitored data has demonstrated the spatially local effects. A correlation between SVF and temperature swing has been derived from the data presented above which, for urban situations with SVFs less than 0.85, is as follows (Figure 6.8):

$$T_{swing} = \ln(1 + SVF/0.02)/0.35 \qquad (6.4)$$

Using computer analysis it is possible to plot the SVFs on a pixel-by-pixel basis, allowing a degree of resolution that is appropriate for studying

6.7
A digital elevation model (DEM) of a part of Cambridge, UK, showing street level as white and the tallest buildings as black

Predicted peak summertime temperature swing as a function of the sky view factor, showing that the greater the view of the sky from the ground, the greater the temperature swing

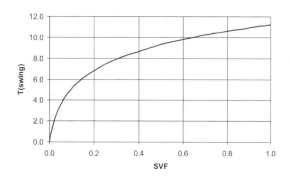

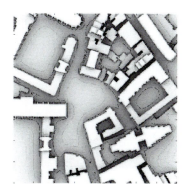

6.9

Sky view factor distribution for a part of Cambridge

6.10

Shadow casting on 22 June for the site in Cambridge

spatial diversity between urban spaces. Figure 6.9 shows the SVFs for a part of Cambridge, where darker areas indicate lower SVFs and thus a lower temperature swing compared to the ambient conditions. This area of Cambridge has an average SVF of 0.7 which, using Oke's correlation (equation (6.1)), creates a maximum heat island intensity (ΔT_{max}) of 5.5°C. However, it is clear that the variation in SVFs across the urban area is significant, ranging from approximately 0.9, equating to a ΔT_{max} of 2.8°C in the most open areas, to 0.2 in the deep courtyard spaces and narrow alleys, where ΔT_{max} is 12.4°C. For that range of SVFs, the range of temperature swings, using equation (6.4), would be between 7 and 11°C, which means that one would, in theory, expect a peak temperature difference of 4°C to be available to the user. This compares with the 6.4°C range measured in Cambridge. The theoretical figure of 4°C excludes intermediate spaces because the digital elevation model (DEM) cannot model open spaces with opaque roofs. Nor does it include an assessment of orientation or scale. A value of 4°C can thus be taken to be significant in terms of thermal comfort, particularly as further diversity resulting from solar radiation and air movement have not yet been taken into account.

Using the DEM of an urban area it is a relatively simple procedure to project shadows for any time of the day and day of the year. Figure 6.10 shows the accumulated shadows for the site in Cambridge for a midsummer day.

From the shadow casting image it is possible to show how many hours of sun any part of the urban area receives. In order to simplify the image one can generate contour maps, establish threshold values and from this define zones of predominantly sunny or shady conditions. Figure 6.11 is an example that shows urban zones receiving less than six hours of sunshine. Because solar radiation has a significant impact on thermal comfort, as discussed earlier, the degree of availability of sun and shade as represented by a threshold value for hours of shade is a simple indicator of spatial diversity.

The final environmental parameter that is discussed is wind. In order to simplify the analysis for the purpose of demonstrating the notion of diversity, we have used the concept of wind shadows created by the sheltering effects of buildings. The length of the wind shadow is a function of the form

6.11
Shadow map of Cambridge, where black zones are those that receive less than 6 hours of sunlight on 22 June when the theoretical maximum is 15 hours

6.12
Cumulative wind shadow casting for the Cambridge site

6.13
Wind shadow map where black areas indicate areas predominantly in wind shadow

and orientation of the building in relation to the wind direction. Because the wind direction changes we have used a wind rose to determine the frequency and direction of the wind in order to inform the relative 'intensity' of the wind shadow. Thus, a wind shadow generated by the prevailing wind direction will be more significant than a wind shadow created from a less significant direction. For this project we use the wind speed/frequency rose for Cambridge in order to map wind shadows (Figure 6.12). These overlapping shadows are then used to determine those areas which are predominantly still compared to those predominantly exposed to wind in order to produce an aggregated yet simplified map (Figure 6.13). It would clearly be possible to produce much more detailed and time-dependent wind analysis, using for example CFD (computational fluid dynamics) analysis. However, in this context our approach is used to demonstrate a methodology for deriving a diversity map that can combine simple thermal, solar and wind parameters.

The final stage in the mapping of environmental diversity is to overlay some or all of the notional maps of thermal, solar or wind conditions. For example, the combination of the solar and wind maps will indicate the relative amount of space that is:

- sunny and still;
- sunny and windy;
- shady and still; and
- shady and windy.

For the Cambridge site on a summer's day Figure 6.14 shows how the combination of these two factors can give a sense of potential environmental diversity in an urban context. The map indicates that an intricate mix of environmental conditions is available. Figure 6.15 shows that the site has a predominance of sunny-still conditions: quite appropriate for the English climate, particularly as there is significant availability of still-shady as well as

6.14
Diversity mapping of sun/shade and windy/still areas

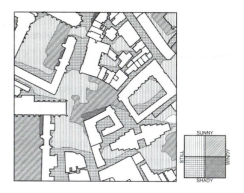

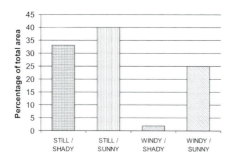

6.15
Relative distribution of environmental characteristics

6.16
Diversity map showing solar, wind and SVF/temperature characteristics of the Cambridge site

windy-sunny areas. For a climate which is temperate, the almost complete lack of windy-shady areas is not of great concern, but on exceptionally hot days, which are predicted to increase as a result of climate change, it may present noticeable limitations on comfort. Remedial measures such as the introduction of shading and evaporative cooling from trees and water features will be a way of improving the situation. Additionally, one could conceive of enabling access to intermediate spaces with the appropriate thermal characteristics such as high thermal mass, orientated to prevailing winds to encourage air movement, northerly aspects, etc.

The final figure shows the mapping of solar, wind and SVF (Figure 6.16). SVF can be expressed as summertime resultant temperature swing and shows the more stable thermal environments (black areas have a lower T_{swing}) compared to those that follow synoptic conditions more closely (higher T_{swing}). In the more stable environments, particularly the deep courtyards, the summer daytime temperatures have been measured as cooler than ambient, confirming the presence of urban cool islands. However, night-time temperatures in such occluded spaces are warmer than the ambient temperature and adhere to the urban heat island intensity as defined by Oke.

7.1
Square in San Sebastian, Spain

Need for open spaces

The concentration of activities in cities provides a necessary and stimulating social, cultural and economic milieu and open spaces offer the backdrop for these activities to take place. Whether these spaces have been shaped by commerce and defence, political systems and cultural traditions or architectural fashions, or other intrinsic factors such as climate and topography, they are necessary for urban life, as they encompass the nature of cities, offering extensive opportunities for communication.

Open spaces in the urban environment support a great range of human needs including relaxation, passive as well as active engagement. The social role of open spaces is particularly evident in working-class areas (Hartman 1972; Jacobs 1961), where external spaces and neighbourhoods are extensively used both for socialising with neighbours and for taking the children out to play, a particularly acute need in cases of limited housing space.

Since quality of life stems from the public realm, it is indispensable to consider public open space as complementary to the built environment, providing comfortable, clean and satisfactory open spaces for the citizens, allowing increased social interaction. Simply increasing the amount of open space given to the city does not ensure social cohesion. In New York, in 1972, developers were given incentive bonuses for providing plazas; for each square metre of open space they would give to the public, they would gain ten square metres of extra floor space over the level normally permitted. As a result, plazas surrounding expensive developments rapidly increased, but, in the vast majority, remained empty of people apart from the occasional passer-by, eventually enhancing isolation and social exclusion.

On the other hand, there is the San Francisco experience, where public interest in the quality of open spaces has led to different referendums for protection of sunlight access in parks and protection from the wind effects caused by new developments (Bosselmann *et al.* 1988). In this framework, the microclimatic conditions of open spaces are of core importance for their success, and the hypothesis examined is that the resulting comfort conditions affect people's behaviour and usage of outdoor spaces.

Outdoor thermal comfort

Conventional comfort theory relies on a steady-state model where the production of heat is equal to the heat losses to the environment, aiming to keep a constant core body temperature of 37°C, so that the environmental conditions, which provide thermal satisfaction, dependent only upon the activity of the subjects and their clothing level, fall within a narrow band. Subsequently, work showed that people take action to improve their comfort conditions by modifying their clothing and metabolic rate, or by interacting with the building — 'adaptive' actions (Nicol 1990) — whereas intrinsic factors proved to be important for thermal satisfaction. The environmental parameters affecting thermal comfort conditions outdoors, even though similar to indoors, are encountered within a much wider range and are more variable. Therefore, due to this complexity, in terms of variability, temporally and spatially, as well as the great range of activities in which people are engaged, there have been very few attempts to understand comfort conditions outside. Within these rare attempts to understand the thermal conditions outdoors, a purely physiological model has been used, similar to the model used for the indoor environment, adapted for the solar radiation parameter. However, there is a lack of under-standing of the human parameter in these spaces, and the subjective responses of humans. If 'adaptive opportunity' is significant enough to call for a change of the comfort standards used by the building services' profession, what would be the situation outdoors, where the environment is uncon-strained and the only modification is action taken by people?

In order to study the effect of the subjective response to the microclimate and the corresponding comfort conditions in the use of open space, an extensive research project was carried out in Cambridge, UK, investigating thermal comfort in the urban context, shedding some light on the complexity of issues involved (Nikolopoulou *et al.* 1998 and 2001). Investigating the comfort conditions of open urban spaces, a differentiation has to be drawn between routes and resting places. In this work we are concerned with resting areas, as people choose to sit somewhere, whereas they do not necessarily choose a particular route, in order to avoid discomfort. Nevertheless, this discomfort will not cause them serious stress, since the time of exposure to the specific environmental conditions is short. On the other hand, with resting places the situation is different, as poor comfort conditions may distress people and lead them to avoid using these areas.

Field surveys were employed for the evaluation of the micro-climate and thermal comfort conditions at four different open areas in Cambridge city centre (Figure 7.2) within the year 1997 to take into account the seasonal variation, which affects the use of space. People were studied in their natural environment through observations and personal interviews, to evaluate their perception of the thermal environment. Objective environmental parameters affecting the thermal environment (air temperature, solar radiation, wind and humidity) were recorded in a portable mini-met station, which was placed next to the interviewee. This information was then compared with subjective human behaviour and responses (from structured interviews and observations) to evaluate the thermal comfort conditions that people experience. Individuals' characteristics and behavioural patterns were also taken into account. The structured interviews, with standard questionnaires, aimed to represent the views of a broad range of users. The thermal aspects of the questionnaire enquired about people's evaluation of the thermal environment and satisfaction, as well as reasons for using the space, frequency of use, etc. A total of 1431 people participated in the field studies.

7.2
Outdoor spaces surveyed in Cambridge, UK

Use of open space

Investigating whether thermal and, by implication, comfort conditions affect people's use of outdoor spaces, through the number of people using the spaces at various intervals, it became apparent that warm conditions and the presence of sunlight are important factors in the use of the space (Figure 7.3), as the average number of people sitting in the space increases as globe temperature increases (Figure 7.4).

Disaggregating the results for the different sites and examining the situation in the site where there are no means of shading available (King's Parade), it is apparent that the curve, although of increasing slope, tends to stabilise after about 25°C. This suggests that, although the sun is much appreciated during the cooler season, in warm conditions fewer people are using the space (Figure 7.5), and due to the general lack of seating areas available in the city centre, people do sit, but for shorter periods than they would otherwise do, e.g. to eat a sandwich, etc.

7.3
Good weather and places for people to sit increase the use of the space dramatically

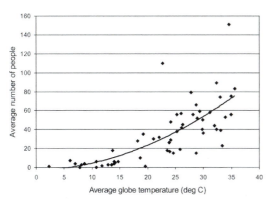

7.4
Average number of people using outdoor spaces as a function of globe temperature (°C), in all the different sites

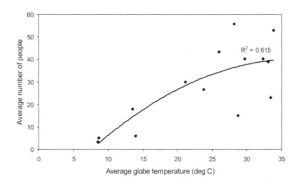

7.5
Average number of people using outdoor spaces as a function of globe temperature (°C), in King's Parade, where there are no means of shading

Subjective comfort conditions

During the field surveys, the interviewees were reporting their thermal sensation and value judgement, on a five-point scale, varying from 'too cold' to 'too hot'. Investigating the effect of microclimatic parameters on people's thermal comfort evaluation revealed that temperature, wind speed and sunshine as the most important parameters of comfort outdoors, influencing thermal sensation, with air temperature being the main determinant of comfort for the specific climatic context (with the expectation that the situation would presumably be different in a humid climate). However, such effects accounted for only 50% of the variation in thermal sensation, which indicates that other parameters become significantly important.

Comparing such subjective data with the thermal index Predicted Mean Vote (PMV) (ISO 7730 1994), taking into account the mean objective environmental parameters recorded for the duration of the interview, clothing levels and metabolic rate, for each interviewee, revealed great discrepancy between the two sets of data, regarding thermal comfort conditions outdoors. The percentage frequency distribution combined for the different seasons (Figure 7.6) demonstrates this discrepancy. More than 50% of the people have voted for the warm part of the scale, around 20% for the cool, whereas for the extreme ±2, the value is around 10%. The corresponding theoretical PMV curve is very different and much flatter, the hatched area representing the difference between the area underneath the PMV lying outside the Actual Sensation Vote (ASV) curve. Only 35% of the interviewees are within the acceptable comfort conditions, the vast majority lying in conditions which are too hot or too cold, since the PMV calculated for the conditions that people have experienced lie from −9 to +7!

To present this as a measure of people's dissatisfaction for the various conditions, the Predicted Percentage of Dissatisfied (PPD), based on the theoretical calculation of the PMV for each interviewee (ISO 7730, 1994) (getting a mean value from the individual calculated PPDs), was compared with the corresponding Actual Percentage of Dissatisfied (APD) (Figure 7.7). The PPD varies from 56% in spring to 91% in winter, whereas the yearly average is 66%. That implies that 944 of the 1431 people sitting outside should be dissatisfied

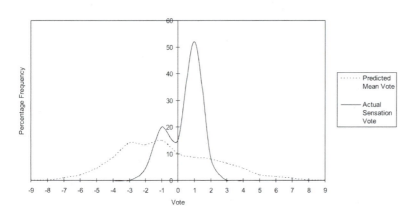

7.6
Percentage frequency distribution of Predicted Mean Vote (PMV) and Actual Sensation Vote (ASV) for all sites and the whole year

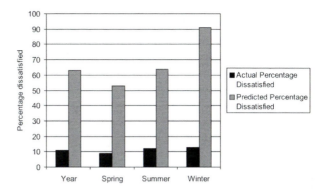

7.7
**Comparison between Actual
Percentage Dissatisfied (APD) and
Predicted Percentage Dissatisfied
(PPD)**

with their thermal environment. In fact, the APD is always around 10%, a figure that is regarded as acceptable, found even in controlled indoor environments. This suggests that adaptation takes place, and a purely physiological approach is inadequate to characterise thermal comfort conditions outdoors, also reinforced by the following figures.

Adaptation

The term 'adaptation' can be broadly defined as the gradual decrease of the organism's response to repeated exposure to a stimulus, involving all the actions that make them better suited to survive in such an environment. In the context of thermal comfort this may involve all the processes that people go through to improve the fit between the environment and their requirements. Within such a framework, adaptive opportunity can be separated into three different categories: physical, physiological and psychological (Nikolopoulou et al. 1999), as explained below.

PHYSICAL ADAPTATION

Physical adaptation involves all the changes a person makes, in order to adjust to the environment, or alter the environment to his or her needs. We can identify two different kinds of adaptation, reactive and interactive. In reactive adaptation, the only changes occurring are personal, such as altering one's clothing levels, posture or even position. In interactive adaptation, however, people make changes to the environment in order to improve their comfort conditions, such as opening a window, turning a thermostat, opening a parasol, etc. Evidence for such forms of adaptation outdoors is described below.

REACTIVE ADAPTATION

Clothing adjustment is a common form of adaptation, helping people to adapt to a wide range of temperatures. Consequently, there is a good fit, between air temperature and clothing insulation worn outdoors (Figure 7.8), as well as between clothing levels and people's thermal sensation vote.

The consumption of cool or hot drinks can be viewed as an adaptive action, affecting the individual's net metabolic heat, decreasing it by 10%

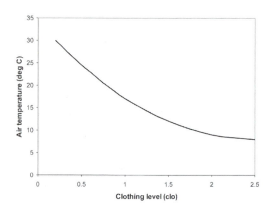

7.8
**Variation of clothing
insulation with air
temperature**

(Baker and Standeven 1996), or increasing it by 5% respectively, and can be the result of the prevailing thermal conditions. Investigating the consumption of cool drinks and the respective climatic conditions, it was interesting to notice that this increased as air temperature increased, suggesting that the consumption of cool drinks may be partly as a result of the thermal environment (Nikolopoulou *et al.* 1999).

Changing position, when possible, is also an effective way of avoiding discomfort. This kind of adaptive opportunity is highly relevant to outdoor spaces, where constraints of workspace, etc., do not exist, therefore people are free to move around as they wish. In summer, spatial variation involved people moving to a shaded place, whereas in winter protection from the wind, or from a damp place was a factor. More specifically from the people interviewed, 43% of the variation of people sitting in shade can be attributed to air temperature alone (Figure 7.9). It is also worth pointing out that for the specific climatic conditions, it was revealed that 25°C seems to be a critical point regarding the transition of people from sun to shade (Figure 7.10), with a mean wind speed of 1 m/s.

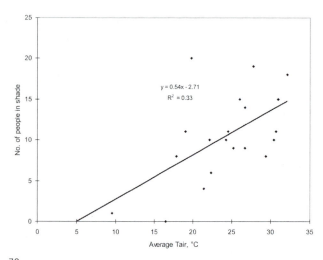

7.9

Variation of people sitting in shade as a function of air temperature

7.10
**At air temperatures above 25 °C people generally
seek out shade**

INTERACTIVE ADAPTATION

Unlike the situation indoors, interactive adaptation outdoors is not frequent in the form of environmental control. Primarily, there are not many elements allowing such interaction with the environment. Second, even in places where features such as parasols were available, no interaction was noticed from the public, but only from a member of staff. This must be due to the fact that people do not feel they have the authority to impose changes in the area, particularly since such an action would affect the neighbouring tables of a restaurant, not wishing to impose on others. This is in agreement with the situation indoors, where it has been found that occupants of private offices exercised more environmental control than occupants of partitioned offices, who in turn exercised more control than the occupants of open plan spaces (Heerwagen and Diamond 1992).

PHYSIOLOGICAL ADAPTATION

Physiological adaptation implies changes in the physiological responses resulting from repeated exposure to a stimulus, leading to a gradually decreased strain from such exposure. In the context of the thermal environment this is called physiological acclimatisation, a crucial mechanism in extreme environments, but not of central importance in the use of outdoor spaces.

EVIDENCE FOR PSYCHOLOGICAL ADAPTATION

Different people perceive the environment in different ways, and the human response to a physical stimulus is not in direct relationship to its magnitude, but depends on the 'information' that people have in a particular situation. Psychological factors therefore influence the thermal perception of a space and the changes occurring in it, as described below (Nikolopoulou and Steemers 2003).

NATURALNESS

This is a term employed by Griffiths *et al.* (1987) describing an environment free from artificiality, whereby there seems to be increasing evidence that people can tolerate wide changes of the physical environment, provided there is a direct link and an association with natural conditions. Perhaps the best example that this is an important parameter in people's perception of outdoor spaces is highlighted in Figure 7.7, where the respective PPD and APD profiles are very different. In outdoor places, where all the climatic changes are perceived and occur naturally, wider changes of the physical environment are tolerated.

EXPECTATIONS

Expectations — that is what the environment should be like, rather than what it actually is — greatly influence people's perceptions; for example, in naturally ventilated buildings people expect variations in temperatures, both temporally and spatially, whereas in air-conditioned spaces they expect a much more stable thermal environment. In outdoor spaces this is supported by the frequent reply that people gave throughout the year 'it's OK for this time of year', 'or for this time of year I'd prefer it warmer'. In the few instances when thermal conditions deviated from what people had experienced during the previous days, this caused differences in people's sensation votes, or even complaints, as their expectations had changed. This is in agreement with the findings of a Norwegian study (Zrudlo 1988), where minimum comfort temperature in autumn was 11°C, whereas in spring it was 9°C. Expectations varied as a result of the much cooler temperatures preceding the spring.

EXPERIENCE

Experience directly affects people's expectations and can be differentiated in the short and long term. Short-term experience is related to memory and seems to be responsible for the changes in people's expectations from one day to the next. This is also the reason that thermal neutrality for outdoor conditions varied from 7.5°C in winter to 27°C, in summer (Nikolopoulou *et al.* 2001), as physical adaptation only partly justifies this range of temperatures. Long-term experience is related to the schemata that people have constructed in their minds, determining a choice of action under different circumstances. Therefore changes in clothing, consumption of cool drinks to alter metabolic heat, moving from sun to shade, etc. all represent well-established choices of action on the issue of how to cope with the variable thermal environment.

TIME OF EXPOSURE

Exposure to discomfort is not viewed negatively if the individual anticipates that it is short-lived, such as getting out of a warm car to enter a building in winter and no significant dissatisfaction is caused. This is a critical factor for external spaces which, apart from movement, are mainly used for recreational activities, and people can modify the time they spend outside, according to their needs.

PERCEIVED CONTROL

It is now widely acknowledged that people who have a high degree of control over their environment tolerate wide variations and thus reduce negative emotional responses. In the surveys, this was apparent in the form of people mentioning choosing their sitting positions so that the option of both sun and shade was open to them, and they could move freely from one to the other if required. It is not important whether they actually moved position in the end, the critical issue was that this choice was available. Another form of perceived control over the environment became apparent when investigating the reasons people gave for being in the spaces, particularly in relation to their comfort state. It became apparent that the number of people feeling uncomfortable and dissatisfied with the thermal environment was higher when the only reason for being there was to meet someone, rather than for other reasons. In particular, 23% of the population using the space as a

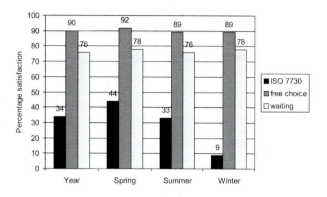

7.11

Percentage satisfaction: (i) theoretical predictions according to ISO 7730 (PPD); (ii) people in the area by their own choice; (iii) those waiting for another person to arrive

meeting place, waiting for another person to arrive, reported dissatisfaction with the thermal environment (Figure 7.11). This amount of dissatisfaction decreases by half to 12%, for the population that have gone to the space for other reasons.

Therefore, people who are in the space for various reasons are aware that it was their own choice to expose themselves to these conditions, and when they wish to they can leave, making them more tolerant to the thermal environment. However, people that were not there by their own choice but to meet someone, did not have the option of leaving when they so wished. The termination of their exposure to the thermal conditions was dependent on external factors, in this case the arrival of the other person, which was causing distress, making them less tolerant of the environment. This issue of free choice becomes of prime importance in outdoor spaces, where actual control over the microclimate is minimal, perceived control having the biggest weighting.

ENVIRONMENTAL STIMULATION

Comfortable conditions have been regarded as those where occupants feel neither warm nor cool, where ambient conditions are 'neutral'. However, it is increasingly believed that a variable, rather than fixed, environment is preferred whereas a static environment becomes intolerable.

Environmental stimulation is an issue of primary importance in external spaces, where the environment presents few thermal constraints, this being an important asset of such areas and one of the reasons that people use these spaces. Whatever people have come outdoors for, they are constantly faced with enormous sensory stimulation whether pleasant or unpleasant. Views change, sounds and smells vary, whereas at the same time the thermal environment changes as one moves from narrow streets to open squares: cool, warm, shaded, exposed, humid, sticky, fresh. Thus the quest for pleasant sensory stimulation is also implicit in all the reasons that people visit open spaces, rather than preferring the more 'sterile' indoor environment. Stimulation was indeed the main reason for the majority of people to sit outdoors, for visiting the area, and what people described as enjoyable about the areas.

7.12
**A stimulating environment increases
tolerance for poor comfort conditions**

This variability and stimulation is especially desirable for people who were working in a building and were coming out for their lunch-break. An interesting point arose by examining the number of people sitting in sun and shade, in spring and summer, in relation to their immediate thermal history — that is where they were before — in a building or outdoors. It was noticed that there is a tendency for people coming out of a building to sit in the sun rather than shade, whereas the majority of people sitting in the shade come from outside. The need to 'charge up' the body with sun was greater than the short-lived and non-threatening physiological strain on the body. On the contrary, the majority of people found sitting in the shade had come from outside.

The most plausible justification seems to be that they see the external environment with the fresh air, the sun and the wind as invigorating stimulation for the senses, wishing to spend some time there before returning to the more monotonous workplace. On the other hand, people who had spent enough time outdoors before arriving in the area had already achieved sensory and thermal equilibrium. This also suggests that heat storage, a term which is not included in the theoretical PMV model (ISO 7730), may also be a significant parameter in the evaluation of thermal comfort.

Therefore, if the site itself is very interesting, offering different kinds of stimulation, people will have higher tolerances to the extreme conditions (Figure 7.12), provided they are not threatening, than they would under average circumstances.

Effect of adaptation

Thermal comfort conditions outdoors allow for adaptive opportunity to be developed to its full extent, both physically and psychologically. This explains

7.13

Schematic diagram of the adaptive opportunity continuum

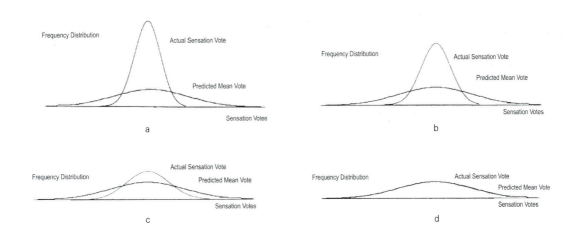

7.14

Hypothetical frequency distribution for ASV and PMV: (a) outdoors — maximum adaptive opportunity is responsible for the great discrepancy between the two votes; (b) in free-running buildings — sufficient adaptive opportunity allows considerable discrepancy between the two votes; (c) in closely controlled HVAC buildings — little adaptive opportunity results in small discrepancy between the two votes; (d) in climate chambers — lack of adaptive opportunity results in the two votes being nearly identical

the great discrepancy found between the ASV and PMV, with 75% of the population falling outside the acceptable comfort conditions as defined by the theoretical physiological model, whereas actual dissatisfaction was always around 10% (Figure 7.7). We can, therefore, consider adaptive opportunity as a continuum (Figure 7.13). At one end of the spectrum, the thermal environment is totally uncontrolled and variable, such as outdoors with adaptation developing fully both physically and psychologically (Fig. 7.14(a)). At the other end, the environment is fully controlled, such as in climate chambers, with no adaptation taking place (Fig. 7.14(d)). Buildings can be allocated at various points in between according to the degree of adaptation they allow for. Closely controlled, fully air-conditioned buildings not allowing interaction between the occupants and the system would be closer to the climate chamber, whereas free-running buildings are closer to the outdoor situation.

With this adaptive opportunity continuum in mind, it should be possible to identify hypothetical relative frequency distributions for ASV and PMV for the different categories presented in Figure 7.6. Figure 7.6 can therefore be redrawn for outdoor spaces on a theoretical basis corresponding to one end of the adaptive opportunity continuum (Figure 7.14(a)). At the other end of the spectrum we have the climate chambers. Since the physiological comfort model was developed in climate chambers, one would expect the frequency distribution for ASV and PMV to be identical (Figure 7.14(d)).

Moving away from the extremes, in the situation where the building envelope is sealed and the climate is fully controlled by a central plant, we would expect the ASV and PMV curves to be very close as a result of minimal adaptive opportunity (Figure 7.14(c)). This expectation is further strengthened by de Dear *et al.* (1997), who explain that the PMV model describes thermal sensations for closely controlled buildings rather well. However, in free-running buildings where a higher degree of adaptation is feasible, the difference between the two curves becomes significant (Figure 7.14(b)).

Figure 7.13 presents the broad categories of different environments in terms of adaptive opportunity. It is reasonable to assume that differences in the degree of adaptation still exist within each of these groups, although of much smaller magnitude. For example, in free-running buildings, the degree of adaptive opportunity can vary between a cellular and an open plan room. This is also the case with outdoor spaces, allowing for a variety of spaces, in terms of sun and shade, etc. Although it is possible to evaluate the effects of physical adaptation with respect to environmental diversity, it is particularly important to understand the parameters that relate to psychological adaptation.

Influence of psychological adaptation

Understanding the interrelationship between the different parameters of psychological adaptation would be of interest in order to compare their relative significance, and to assess their design role, that is whether design considerations would influence these parameters or, vice versa, whether they could influence design decisions. It is therefore possible to present a speculative network of the different parameters affecting psychological adaptation with lines of influence between them, indicating whether one parameter affects another (Nikolopoulou and Steemers 2003). In a simplified way this can be presented graphically (Figure 7.15), indicating the degree to which one variable influences and is being influenced by another. The three variables that are influenced by all other parameters of psychological adaptation are presented in one group, to denote the interrelations between

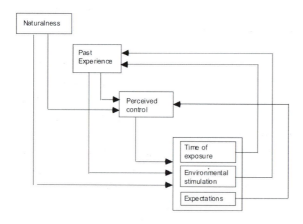

7.15
Network demonstrating interrelationships between the different parameters of psychological adaptation

them, with external connections provided where required. Clearly, this implies that the relationships between those variables are complex, as it is not a simple cause-and-effect situation. Satisfaction with the thermal environment of the space will depend as much on the space itself, as it will on the personal variables that people bring to the area with them, and the former will affect the latter, whereas the latter will affect the perception of the former.

Design considerations

Despite the complexity of the above interrelations, it is possible to consider — without being deterministic — design issues, which would have some impact on the above parameters. This would, in turn, increase the range of psychological adaptation that could take place, therefore widening the range of environmental conditions considered as comfortable. An awareness of these issues would be valuable to architects, planners and urban designers, not by way of limiting possible solutions, rather by enriching the design possibilities.

The first parameter considered, influencing without being influenced by other parameters, is naturalness, which is part of the character of a place. This can be significantly increased, by 'greening' an area, adding vegetation or views of landscape, particularly within the dense urban context, which would accentuate the distinct character of the different areas.

Past experience, although it is a variable that people bring to the space rather than being characteristic of the space, is still able to affect the short-term experience, mainly in the context of the design of the urban fabric and urban block. Since people's thermal sensation is influenced by their immediate short-term thermal experience, by providing more spatial variety in the city, a rich variety of different environments can be experienced, between indoors and outdoors, affecting their thermal sensation (discussed in detail in Chapter 6 by Steemers *et al.*).

Perceived control can be affected by providing increased opportunities for physical adaptation to take place. Regarding reactive adaptation, spatial variation can be site-related by providing a variety of sub-spaces within the same area. This would translate into allowing access to the sun as well as to shade, exposure to breezes as well as protection from the wind, with different areas being preferred in different seasons. Microclimatic control is feasible with the use of vegetation for shading and wind breaks, movable canvas awnings, canopies made of various materials such as reeds, bamboo, or vines, etc. Similarly, transition spaces such as arcades provide useful spatial variation in areas that have to cope with harsh winters. As for interactive adaptation, although infrequent in outdoor spaces, movable elements such as parasols or awnings (Figure 7.16), provide environmental control with protection from sun and rain, and are normally appreciated by users of the space.

The role of the architect is therefore of major significance, determining the character of the area. An exemplary approach for the shading of large open spaces can be seen in Bobo Rasch's solution for a historic mosque in Madinah, Saudi Arabia, which is sensitive to the existing urban fabric. Six large fabric umbrellas were designed, each 17 × 18 m^2, 14 m high, standing on steel columns, all creating a convertible roof without undermining the character of the mosque (RIBA, 1997). The umbrellas when open cover the entire courtyard, whereas at night they close to allow the hot air to be dissipated into the sky.

The abovementioned parameters can be influenced from the design point of view, affecting the remaining time of exposure, environmental stimulation and expectations. Time of exposure is a personal variable, but may be influenced by people's thermal evaluation of the area, positively for extending their stay in the area, or negatively reducing it.

Regarding the degree of environmental stimulation desired, although determined by the individuals, protection from negative aspects and exposure to positive aspects of the climate can increase such desire

7.16
Diurnal adaptation of shading canopy

7.17
Combining a stimulating environment with evaporative cooling

(Figure 7.17). Microclimatic planning can thus increase the use of outdoor spaces during the intermediate seasons. In fact, in Norway, it was found that the outdoor season could be extended by up to six weeks during the more critical seasons of spring and autumn by appropriate microclimatic planning. This meant providing protection from the wind, orientation to maximise solar exposure, avoiding overshadowing, employing heat-absorbing and heat-reflecting materials, etc. (Culjat and Erskine 1988).

Finally, expectations are also linked to the design of open spaces indirectly, by affecting the degree of perceived control.

Conclusions

This work has thrown some light on the complexity of issues involved in thermal comfort and diversity in outdoor urban spaces, particularly in areas identified as resting places, as opposed to routes. It has been revealed that microclimatic parameters strongly influence thermal sensations, as well as the use of open urban spaces throughout the year. Therefore, climate should be taken into consideration at the intermediate scale of the urban block (a scale which has received little attention in research), integral to user satisfaction and therefore to the success of the space. However, such an approach only accounts for around 50% of the variation of the interviewees' actual thermal sensation votes. The rest cannot be measured by physical parameters, but psychological adaptation seems to become increasingly important, accommodating wide fluctuations in the physical environment, so that thermal discomfort is avoided.

The parameters comprising physical and psychological adaptation have been evaluated and their design role has been assessed, in order to increase diversity and thus the occurrence of both kinds of adaptation through the design of urban spaces, to improve thermal comfort and consequently increase the use of these spaces. A parallel can be drawn here between indoor and outdoor environment. When the building fabric is carefully designed, there may be various interactions between the individual and the external fabric, such as windows, shading devices, etc. to increase diversity both spatially and temporally, whereas a poor design response would limit such opportunities.

It is clear that it is inadequate to design open spaces with regard to thermal comfort solely on the basis of a physical model. The physical environment is important in outdoor thermal comfort but psychological adaptation is also an important factor. Although these include largely personal parameters, appropriate microclimatic diversity and careful design of open spaces can provide protection from negative aspects and exposure to positive aspects of the climate, therefore increasing the use of outdoor space throughout the year. Different seasons require different approaches, but a variety of spaces providing different environments would increase both physical and psychological adaptation. An architectural vocabulary that responds to the local climate is more likely to avoid creating harsh micro-climates; windier, hotter or colder streets and squares.

Physical and psychological adaptation are complementary rather than contradictory, and consideration of this duality could assist in designing diverse conditions to encourage and increase the use of the city's open spaces, strengthening social interaction between citizens. Such an approach would be valuable to urban designers, not by way of limiting possible solutions, rather by enriching the design considerations.

Note

Parts of this work were supported by the RIBA Trust Research Award (1996) and awarded International Society of Biometeorology ISB Human Biometeorology Scientific Award (2001). The findings of this project led to the development of a large-scale research project, investigating urban comfort conditions across Europe, involving twelve organisations, completed in 2004. The project, entitled 'RUROS: Rediscovering the Urban Realm and Outdoor Spaces' (funded by the EU 5th Framework Programme, City of Tomorrow and Cultural Heritage from the programme Energy Environment and Sustainable Development), is coordinated by Dr Marialena Nikolopoulou at CRES, and Dr Koen Steemers at the Martin Centre is one of the principal partners.

References

Baker, N. and Standeven, M. (1996) 'Thermal comfort for free-running buildings', *Energy and Buildings*, 23: 175–82.

Bosselmann, P., Dake, K., Fountain, M., Kraus, L., Lin, K. T. and Harris, A. (1988) *Sun, Wind and Comfort: A Field Study of Thermal Comfort in San Francisco*. Working paper 627, September, Centre for Environmental Design Research, University of California Berkeley.

Culjat, B. and Erskine, R. (1988) 'Climate-responsive social space: a Scandinavian perspective' in Mänty, J. and Pressman, N. (eds) *Cities Designed for Winter*, Helsinki: Building Book Ltd.

de Dear, R., Brager, G. and Cooper, D. (1997) *Developing an Adaptive Model of Thermal Comfort and Preference*, Final Report on ASHRAE RP-884, Sydney: Macquarie University.

Griffiths, I. D., Huber, J. W. and Baillie, A. P. (1987) 'Integrating the environment' in Steemers, T. C. and Palz, W. (eds) *Proceedings of the 1987, European Conference on Architecture*, Dordrecht: Kluwer Academic Publishers.

Hartman, C. W. (1972) 'Social values and housing orientations' in G. Bell and J. Tyrwhitt (eds) *Human Identity in the Urban Environment*, Middlesex: Penguin.

Heerwagen, J. and Diamond, R. C. (1992) 'Adaptations and coping: occupant response to discomfort in energy efficient buildings', *Proceedings ACEEE 1992 Summer Study on Energy Efficiency in Buildings*, 1083–90.

ISO (1994) *Moderate Thermal Environments: Determination of the PMV and PPD indices and specification of the conditions for thermal comfort*, ISO 7730, Geneva: International Standards Organization.

Jacobs, J. (1961) *The Death and Life of Great American Cities*, Middlesex: Penguin Books.

Nicol, J. F. (1990) 'Passive buildings need active occupants', in *Proc. World Renewable Energy Congress*, Reading, UK.

Nikolopoulou, M., Baker, N. and Steemers, K. (2001) 'Thermal comfort in outdoor urban spaces: the human parameter', *Solar Energy*, 70(3): 227–35.

Nikolopoulou, M., Baker, N. and Steemers, K. (1998) 'Thermal comfort in outdoor urban spaces' in Maldonando, E. and Yannas, S. (eds) *Proc. PLEA 1998: Environmentally Friendly Cities, Lisbon*, London: James & James.

Nikolopoulou, M., Baker, N. and Steemers, K. (1999) 'Thermal comfort in urban spaces: different forms of adaptation' in Perez Latoore, M. (ed.) *Proc. REBUILD 1999: Shaping Our Cities for the 21st Century*, Barcelona.

Nikolopoulou, M. and Steemers, K. (2003) 'Thermal comfort and psychological adaptation as a guide for designing urban spaces', *Energy and Buildings*, 35(1): 95–101.

RIBA (1997) *Shade*. London: Royal Institute of British Architects and Health Education Authority.

Zrudlo, L. (1988) 'The design of climate-adapted arctic settlements' in. Mänty, J. and Pressman, N. (eds) *Cities Designed for Winter*, Helsinki: Building Book Ltd.

Chapter 8

Intermediate environments

André Potvin

Introduction

The process of mechanisation of architecture and movement, which began with the Industrial Revolution, has reduced the degree of environmental diversity in architecture and urban environments. The quest for internal/external continuity is at the centre of urban history and, in particular, the problem of public versus private spaces. The recent retreat of public space before speculative forces has created an unprecedented interiorisation of the public realm and the loss of urban permeability and diversity. In the name of densification and environmental improvement, the public domain has been controlled and stripped of its spatial and thermal diversity. And yet, intermediate environments such as the arcade represent a unique environmental answer to the densification imperative. Set in the heart of the urban block, it has nevertheless maintained a reference to the street and increased the permeability of the urban fabric (Lemoine 1983). Such semi-public or intermediate spaces act as an interface between public and private realms and provide a sense of continuous diversity. Transitional spaces, which are neither internal nor external spaces, do not call upon artificial means of thermal control and somehow lessen the comfort expectations of the users. These spaces simultaneously engage the thermal sensations and movement of the entire body, and provide a progressive adaptation to a new environment (Potvin and Hawkes, 1994). Whereas environmental determinism creates uniformity, environmental diversity increases the morphological possibilities of architecture and urban form. A rich environmental diversity can therefore be achieved within the variables of architecture by subtle articulations between interior and exterior environments. Transitional spaces such as passages, courtyards, and especially arcades, have proved to be efficient means of providing this continuous diversity by favouring environmental diversity and progressive adaptation of the body to avoid any discomfort that may arise through abrupt environmental transients.

This chapter presents a methodology for the assessment of urban transitional spaces' microclimate such as courts, passages, arcades and atria. A theoretical investigation of environmental transients through the manipulation of Humphrey's thermal comfort equation in outdoor spaces highlights the possible extent and the corresponding bodily adaptation (difficult, conscious or 'subliminal') to thermal transients according to combinations of wind and sun. A thermal transients graph is presented to depict thermal transients according to operative temperature and air movement conditions. Practically, environmental transients as experienced dynamically by a pedestrian when moving through transitional spaces are assessed using a portable sensor array and presented graphically on the thermal transients graph. Results of a more extensive survey of existing arcades in Cardiff city centre are also presented (Potvin 1996).

On environmental perception

The experience of space in architecture is dynamic with periodic or constant movement between areas of a building or between inside and outside. Figure 8.1 illustrates that when moving from a space where the environmental condition is inadequate to a space where it is greater in intensity, a sensation of comfort is felt. If, in successive steps or in a steep change, the stimuli increase too much, the positive sensation wanes and becomes negative. An occupant is thus momentarily conscious of a positive change in environmental conditions, followed by a neutral step where the comfort range is attained and then a negative change.

Two users will therefore judge the conditions in a space differently if they experience the sequence in opposite directions, the subjective judgements of environmental conditions always being affected by the preceding environmental condition or reference level. When the environmental conditions change very slowly, below the threshold of sensation, the change may be subliminal. The body mechanisms tend to adapt to accommodate each successive change, making the subjective effect almost imperceptible.

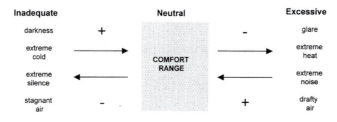

8.1
Environmental transitions (after Flynn *et al.* 1992)

This phenomenon of adaptation occurs both in the increasing and decreasing of a stimulus. Thus, when environmental continuity is required for thermal comfort, changes in the intensity of different environmental stimuli should be subliminal to avoid the sensation of discomfort.

Gibson (1966) first recognised that all five senses are working together in actively seeking information but he postulated that the basic-orienting and haptic systems are particularly relevant to the perception of the third dimension since they encompass the entire body. By placing the entire body at the centre of the perceptual experience, he affirms that no other sense deals as directly with space as the haptic-orienting system, engaging simultaneously feelings of temperature and movement. Thus, the movement between urban spaces of different thermal conditions must have a profound effect on our perception and appreciation of the environment.

Thermal transitions

Few researchers have studied the dynamic impact of environmental variation on the sensation of comfort. Knudsen and Fanger (1990) investigated the impact of temperature stepchanges on thermal comfort in a climate chamber. This study demonstrated that there is a greater sensitivity to cold steps than to warm steps. Moreover, it also highlighted that the speed of adaptation to a new environment depends on whether the step-change is directed away or towards neutral conditions. For the two step-changes away from neutral, the acceptability decreases to below the steady-state level, and at least 20 minutes is needed before the votes reach the steady-state level. This is in contrast to the step-changes towards neutral, where the steady-state level of acceptability is reached within five minutes. In both cases, the immediacy of the thermal sensation response to temperature step-changes supported the hypothesis that it is the rate of change of the skin temperature rather than the actual skin temperature that is responsible for thermal sensation during fast thermal transients in the environment. Knudsen concludes that behavioural thermoregulation may be more important for survival in cold environments than in warm environments. Knudsen's conclusions suggest that it might be more important to prevent important cold steps when moving from one space to another than warm steps, since the former is more likely to trigger a behavioural response than the latter. Extreme behavioural response, such as increasing the metabolic rate by faster movement, may have serious effects on the perception of space, social behaviour, and the overall environmental satisfaction.

At the scale of a building, environmental diversity can be quite easily predicted with computer modelling. However, at the urban scale, the interweaving of spaces is much more complex and difficult to model due to an infinite number of variables.

Speculating on urban environmental transitions

Wind and solar radiation constitute the two main factors affecting comfort in exterior spaces. An environmental transition appears when the combination of these factors varies over time. The variation can be abrupt, creating a marked sense of comfort or discomfort. It can also be subliminal, inducing a sense of neutrality and continuity caused by the slow adaptation of the body to new environmental conditions. A subliminal adaptation is preferable when moving from a comfortable environment into a less comfortable one. In such a case, the subliminal adaptation mitigates a marked sensation of discomfort that would need an important change to revert to the neutral conditions. However, when moving from an uncomfortable to a more comfortable environment, an abrupt transition would appear acceptable. The possibilities of such environmental transitions are infinite and depend as much on the nature of a given climate than on the nature of the built form itself. Table 8.1 shows different combinations of solar radiation and wind exposures. These codes do not refer to any precise values, but they are used to draw a synthetic portrait of possible environmental transients. Combination (02), for instance, signifies an environmental condition characterised by low wind and high solar radiation exposure.

These nine theoretical conditions can be paired to generate 36 different hypothetical environmental transitions such as (01–21) that would signify the passage from a wind-protected, with low solar radiation space to a wind-exposed, with high solar radiation space. The most drastic environmental transition (02–20) would unmistakably be related to the passage from a windy and shaded space to a wind-protected and sunbathed one. However, such a conclusion is not as straightforward for the remaining environmental transitions since wind and solar radiation interact in a very complex way. Here, Penwarden's equation (1973) may help in quantifying the corresponding thermal increment or decrement associated with each of them.

Penwarden adapted a mathematical expression previously developed for internal environments, by adding the solar radiation factor. Equation (8.1) is Penwarden's thermal comfort equation for outdoor activity, which integrates air movement, temperature, solar radiation, metabolic rate and clothing level in the calculation of thermal comfort:

$$Tb - Ta = (M/A_{Du})Rb + k(M/A_{Du})Rc + [k(M/A_{Du}) + S] (4.2 + 13U^{0.5})^{-1}$$

(8.1)

Table 8.1 **Combinations of wind and sun conditions**

		Solar radiation		
		low (0)	medium (1)	high (2)
	low (0)	(00)	(01)	(02)
Wind exposure	medium (1)	(10)	(11)	(12)
	high (2)	(20)	(21)	(22)

where:

T_b = body temperature (37°C)

T_a = ambient temperature (°C)

M/A_{Du} = metabolic rate

k = dissipated metabolic heat by means other than evaporation
= 0.8

R_b = thermal resistance of the body (mean value = 0.065 m^2 °C/W)

R_c = clothing value

S = solar gain per body square meter

U = air movement (m/s)

$(4.2 + 13U^{0.5})^{-1}$ = thermal resistance between clothing and environment (m^2 °C/W).

Equivalent increases or decreases in ambient temperature in order to maintain the state of thermal equilibrium are calculated from equation (8.1). Table 8.2 summarises these equivalent temperatures as a function of the 36 previously discussed hypothetical transitions. The values indicate the passage from A to B whereas moving from B to A would simply result in reversed values.

These equivalent temperatures are interesting indicators of the degree of abruptness of the environmental transition and the corresponding effort needed from the body to overcome the change. Therefore, extreme condition (02–20) corresponds to a 13°C decrease in ambient temperature. Figure 8.2 shows a summary graph of the equivalent temperature differentials for the 36 theoretical environmental transitions.

This graph shows that the equivalent temperature differential varies widely depending on the type of environmental transition. It distinctly identifies three groups of environmental transitions. It is tempting to

Table 8.2 **Equivalent temperature differentials when moving from environment A to B as a function of theoretical environmental transitions**

		(00)	(01)	(02)	(10)	(11)	(12)	(20)	(21)	(22)
						B				
	(00)	0.0	2.5	7.5	-4.5	-3.5	-1.5	-5.5	-5.0	-3.0
	(01)			5.0	-7.0	-6.0	-4.0	-8.0	-7.5	-5.5
	(02)				-12.0	-11.0	-9.0	-13.0	-12.5	-10.5
	(10)					1.0	3.0	-1.0	-0.5	1.5
A	(11)						2.0	-2.0	-1.5	0.5
	(12)							-4.0	-3.5	-1.5
	(20)								0.5	2.5
	(21)									-3.0
	(22)									0.0

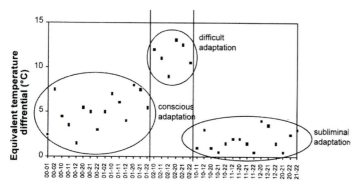

Environmental transition

speculate that these groups should favour subliminal, conscious or difficult adaptations to the new stimuli. A high temperature differential does not, however, imply poor comfort conditions. It simply suggests that the means to achieve thermal equilibrium would be more drastic, such as modification in clothing level or a significant increase in metabolic rate.

Interesting generalisations can be drawn from Figure 8.2 as follows.

1   Passing from a wind-protected and sunlit space to any other space would create on average an important ±11.3°C equivalent temperature differential. It would therefore require more important adaptive behaviour to overcome the transition.
2   Passing from a wind-protected shaded or semi-shaded space to any other would create on average a ±5.1°C equivalent temperature differential. This would be easily overcome by moderate adaptive behaviour.
3   For other transitions, passing from a moderately windy and shaded, semi-shaded or fully sunlit space, or from a wind-exposed and semi-shaded or fully sunlit space to any other, would create on average a ±1.9°C equivalent temperature differential. This would barely be noticed and would therefore be qualified as subliminal.

Thermal transient graph

The quality of the intermediate thermal transient provided by transitional spaces is assessed using a thermal transient graph drawn after Penwarden's thermal equation (8.1) for outdoor activity that still stands as the most complete index of outdoor thermal comfort. By varying the body tissue resistance from the onset of sweating to the onset of shivering, graphs of optimal comfort are established for specific solar radiation and clothing values as a function of ambient temperature and wind speed. However, the present research implies the frequent movement from shaded to sunlit positions, from outdoor to semi-enclosed spaces. In order to gather the

8.2

Graph of the equivalent temperature differentials as a function of environmental transition patterns and the respective adaptation necessary to overcome the thermal transient

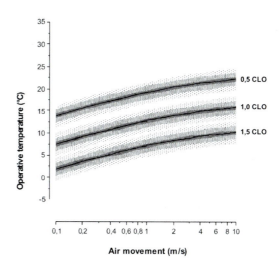

results of the survey on a single, more exhaustive graph, some modifications to Penwarden's original format are necessary.

1   The solar radiation value is zeroed and the operative temperature (mean value of the surveyed ambient and radiant temperatures) is used in the equation instead of ambient temperature.
2   The wide comfort zone is subdivided into several zones corresponding to the common rating of thermal satisfaction ranging from hot to cold according to intermediate body tissue resistance values.
3   The horizontal axis has a logarithmic scale that allows a better distribution of very low values of air movement in semi-enclosed spaces compared to high values in outdoor ones. It is also a more representative graphical illustration of the more important effect of air movement at low speeds on comfort. The length of a line drawn between two points could therefore represent graphically the degree of thermal transition and therefore of the degree or abruptness of the environmental diversity of a particular sequence.

Figure 8.3 illustrates the resulting graph of optimal comfort for variable clothing values as a function of operative temperature and air movement.

The comfort zones drawn on this graph are not immutable since the subjective response of people to their outdoor conditions is not well known. They nevertheless provide a common basis for comparison of several different urban transitional configurations.

Survey method

Recent research in the field of urban environmental sciences has mainly been concerned with the energy performance of cities. Although several researchers have attempted and still work on the prediction of thermal comfort in an urban context, no research has actually attempted to measure the extent of environmental stimuli on the individual. Pignolet (1994), in his recent calculation code for the thermal behaviour prediction of urban spaces, has developed interesting prediction models, but the survey of real conditions constitutes the last step of the study, as a confirmation of the validity of the model. In the present study, surveys come first in order to acknowledge the complexity of the city. The results of the survey are not intended to validate any theoretical model but to initiate a discussion on the importance of urban environmental diversity and transients.

The nature of this study has necessitated the development of an original method of survey to assess the instantaneous environmental stimuli put on the pedestrian by the built form. The survey method, therefore, does not depend on a 24-hour recording of the thermal behaviour of a given space but on the instantaneous effect of its form on individuals as they move through it. The surveys are therefore mostly limited to the period of early

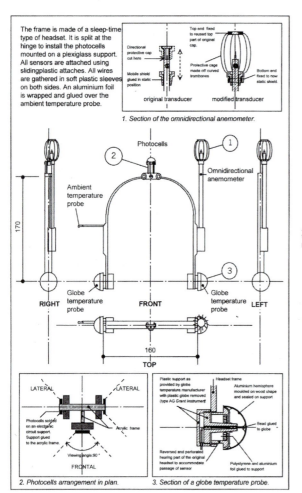

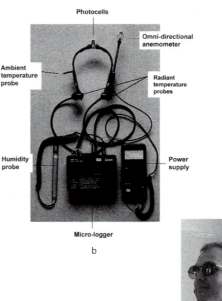

8.4

(a) Construction detail of the portable sensor array (after Baker and Standeven 1994); (b) headset and micrologger; (c) array as worn

afternoon since it constitutes the most critical period of the day during which cities are most occupied and alive with daily activities.

The construction of a portable sensor array was inspired by similar equipment devised for an EC project on indoor thermal comfort by Baker and Standeven (1994). Figure 8.4 shows the construction drawing of the array, made out of a standard sleep-time headset, worn on the head in a similar fashion to a 'Walkman' personal stereo headset, which frees the hands of the surveyor for the performance of other survey activities. The present array incorporates ambient and radiant temperature probes, supplemented with an omnidirectional anemometer, all of which are calibrated for outdoor conditions. Photovoltaic cells are also part of the portable array as the luminous environment provides an idea of the visual transitions according to thermal transients.

This array allows for the assessment of the variation of the thermal environment as experienced dynamically by a pedestrian strolling in the architecture of the city. A hard copy of the data provides a vivid history of the pedestrian's thermal stimulus according to the urban space.

Survey types

The portable array allows for static as well as dynamic surveys. Both types of survey offer particular advantages depending on the intended purpose of the data. In a dynamic survey, the surveyor moves constantly, so as to recreate pedestrian movement in the urban realm. The micrologger remains set to a five-second recording interval, but for an indefinite period. Dynamic surveys offer clear evidence of the extent of thermal differentials to which a pedestrian is exposed when strolling in the city. In this type of recording, illuminance levels are of utmost utility since they clearly record the subtle or abrupt transitions from sunlit to shade, or any other luminous variations.

For a dynamic survey, the instantaneity of the probes' reaction to climatic variations was a major concern since the physiological sensation of warmth by the nerve ends, especially from direct radiation, is instantaneous. Alterations carried out to the previous version of the radiant temperature probes ensured a shorter time response. The slow time response does not affect the validity of the readings, as long as a proper period of time is allowed for the probes to adapt to the variation.

Results

Cardiff city centre offers a great diversity of urban transitional spaces such as public, semi-public, air-conditioned and free-running spaces. A stroll was determined to encompass the maximum of these spaces to conduct some seasonal dynamic surveys using the portable array. Figure 8.5 illustrates

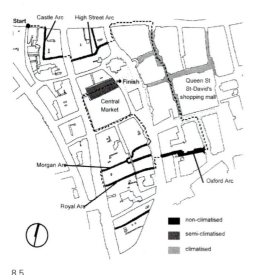

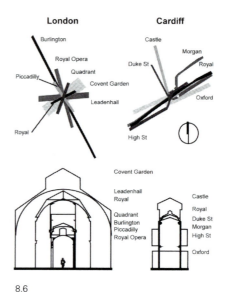

8.5
A stroll in Cardiff city centre

8.6
Comparative plans and sections of arcades in London and Cardiff

Cardiff city centre and the position of major arcades and itinerary of the dynamic survey that starts on Castle Street and ends on St John Street.

The peculiarity of Cardiff's late Victorian arcades does not reside in their grandeur, but in the intricacy of their plans and the way they have developed into a very complex and convivial network of semi-public spaces within the core of the city. Their irregular configurations (Figure 8.6) offer accessibility to spaces set back at a considerable distance from the street. London's arcades are, in comparison, more pompous and exclusive, and are somehow unrelated to the main pedestrian routes.

A cold sunny winter day

Figure 8.7 depicts the environmental stimuli in terms of temperature, air movement and illumination for a stroll in Cardiff's network of intermediate spaces on a cold sunny winter day, under bright sunshine. The thermal gradient graph (Figure 8.8) shows much diversity according to the presence or absence of direct sun radiation. Figure 8.8 demonstrates that the reduction of air movement in semi-public spaces, such as the arcades, manifestly compensates for the absence of direct solar radiation as found in the adjoining open public spaces. In fact, the only outdoor public location to approach comfort conditions is where the pedestrian stands in full sun with protection against the wind. Other semi-public spaces, such as the market and the shopping mall, performed better but they are, however, semi-conditioned. Interestingly, even the thermal conditions of the non-conditioned arcades fall into the comfort zone, affording a lesser clothing level.

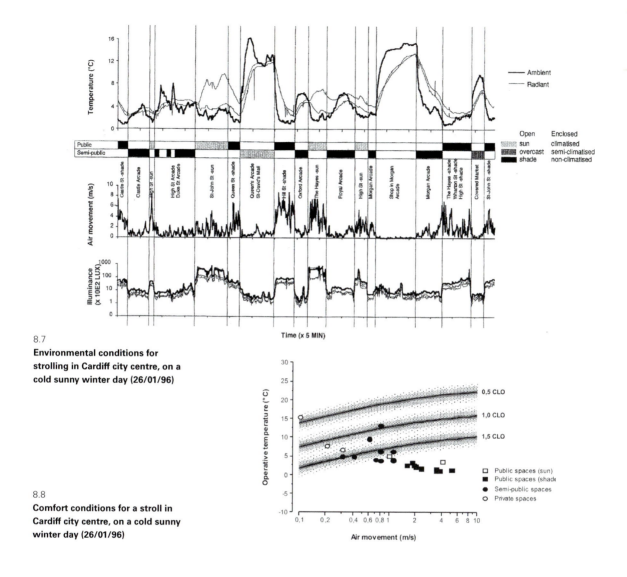

8.7

Environmental conditions for strolling in Cardiff city centre, on a cold sunny winter day (26/01/96)

8.8

Comfort conditions for a stroll in Cardiff city centre, on a cold sunny winter day (26/01/96)

A hot sunny summer day

On a hot sunny summer, the variability of the environmental conditions is maximal as illustrated in Figures 8.9 and 8.10. The high summer sun reaches many urban spaces, creating a visual and thermal chiaroscuro depending on sunlit or shaded spaces. Even the distinction between mechanically controlled and non-mechanically controlled spaces becomes less marked.

Figure 8.10 shows that under such circumstances, the arcade environment lies between the conditions of the shaded and sunlit streets. The direction of the pedestrian plays a major role in the sensation of comfort. When entering the arcade from a sunlit street, the operative temperature suddenly drops by 3°C with a gradual 54% reduction in air movement, which counterbalances the positive cooling effect of the temperature drop. When leaving the arcade towards a shaded street, another 3°C decrease in operative temperature would be felt with an accompanying 46% increase in air

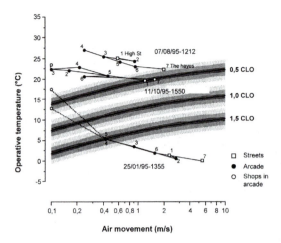

8.12
Comfort conditions for seasonal strolls through Royal Arcade

the only space to create a temporary cooling effect owing to the low and dark passage. With its 100% glazed roof and east–west orientation, Royal Arcade is much exposed to the high altitude summer sun in the early morning and late afternoon.

Morgan Arcade

The architect Edwin Seward began the construction of Morgan Arcade (Figure 8.13) in 1879. Successive Ordnance Surveys maps show that the complex 'Y' shape of the arcade was developed over a period of time, the side crossings being treated as back streets with a pitched glazed roof leading south to the existing Royal Arcade and north to Wharton Street. An unusual central island is surrounded by the glazed roof of the arcade, and is lit by sunlight. Morgan Arcade compares to Royal Arcade in terms of geometry but has a higher opening to volume ratio due to a low, open balustrade that runs under the entire length of the fully glazed roof.

On a bitterly cold winter day, entering the arcade from the wind-battered Hayes Street produces a progressive 2°C temperature increase, while coming from the wind-sheltered High Street there is a significant temperature differential, the temperature of the entrance passage being slightly lower than the street conditions (Figure 8.14). The wind conditions inside the arcade are rather unstable, probably because of the largely open balustrade at roof level. On the other hand, on a warm autumn day, there is no overheating, which is probably owing to the efficient natural convection through the same open balustrade. On a comfortable summer day, there is an important operative temperature differential between the entrance streets, with progressive temperature and air movement transients respectively positive and negative depending on the direction of movement. Position 2, which corresponds to the entrance passage from The Hayes, provides a significant cooling effect by 1°C and a 33% air movement acceleration. This 'C' like pattern of the environmental conditions expresses gradual environ-

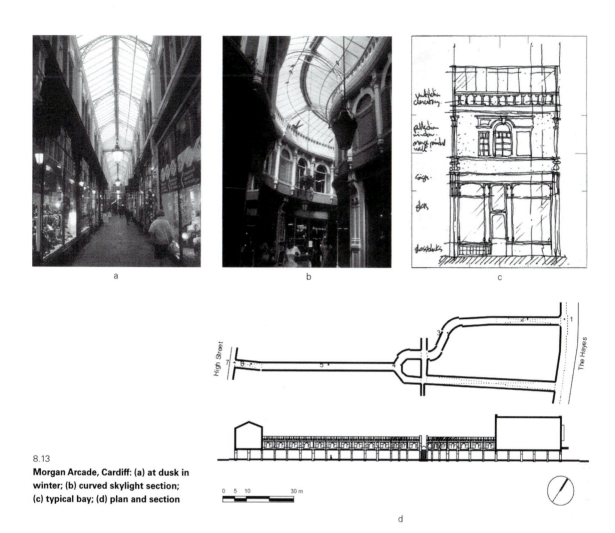

8.13

Morgan Arcade, Cardiff: (a) at dusk in winter; (b) curved skylight section; (c) typical bay; (d) plan and section

8.14

Comfort conditions for seasonal strolls through Morgan Arcade

mental transients and therefore continuity and possible subliminal adaptation by the user. The generous open balustrade at roof level eliminates the inevitable overheating by natural convection in summer, but it should ideally be movable to control the undesirable heat loss in winter. The open balustrade of Morgan Arcade, built some 20 years after Royal Arcade may have been designed in response to the summer overheating of the latter.

Castle Arcade

The architect Peter Rice designed Castle Arcade (Figure 8.15) some 29 years after his design for Cardiff's Royal Arcade. It forms an 'L' shape, linking two busy commercial streets. One-third of its length on the longest axis is two

a

b

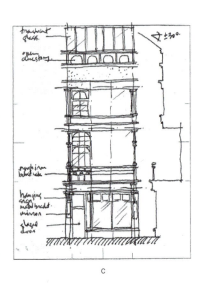

c

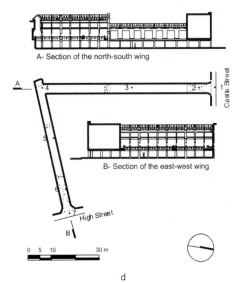

A- Section of the north-south wing

B- Section of the east-west wing

d

8.15
Castle Arcade, Cardiff:
(a) north–south section at noon in
winter; (b) east–west section at noon
in winter; (c) typical bay; (d) plan and
section

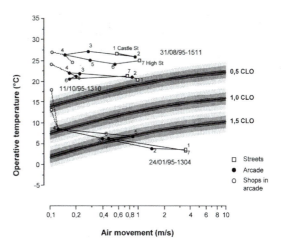

8.16

Comfort conditions for seasonal strolls through Castle Arcade

storeys high with a simple pitched fully glazed roof. The remaining space is three storeys high with a more elaborated pitched glazed roof, supported by open spandrel iron arches. The first floor inner facade of the arcade is recessed to accommodate a continuous balcony joined by footbridges across the arcade, and accessible by lateral open staircases.

The two-storey high section, closed to the outside apart from its entrance passage on Castle Street, counterbalances the generous clerestory openings of the 'L' shaped three-storey high section. It lowers the overall opening to volume ratio of Castle Arcade to 8.2, and ranks it as the fourth most conservative arcade of the corpus of study.

On a bitterly cold and windy winter day, a 5°C progressive temperature increase is recorded when walking from either entrance (Figure 8.16). Air turbulence appears more significant in the highest section, probably owing to the important clerestory openings. This arcade provides a particularly interesting thermal progression from the open street to the protected shops. It allows the temperature in non-heated shops to reach as high as 13°C on the particular cold and windy day of the survey. On a comfortable autumn day, the temperature differential between the streets and the arcade is minimal but air movement is reduced by 73%. On a comfortable summer day, the arcade does not overheat compared to the surrounding streets, and air movement is progressively reduced. It is even locally accelerated in the entrance passage on Castle Street, as also seen at Morgan Arcade, favouring summer comfort by convection.

In summary, these environmental transients analyses have shown that, especially under sunny conditions, ambient environments at opposite entrances of arcades can differ significantly in terms of operative temperature and wind speed. This asymmetrical ambience, resulting mainly from the sun path, was shown to play a major role in the thermal adaptation process, being either positive or negative according to the direction of movement of the pedestrian. Long and narrow arcades provide most progressive and sustained thermal transients, whereas wide or short arcades produce rather abrupt and ephemeral ones. Low and narrow passages at entrances of most arcades produce a gentle cooling effect by wind acceleration in summer.

Environmental transients

Results show the measured operative temperature and air movement of the streets, arcades and their shops for the entire corpus of study for the summer and winter surveys. These graphs, which disregard the many variables of such surveys, nonetheless depict some valuable seasonal tendencies. The record-breaking summer of 1995 proved ideal to investigate the thermal behaviour of arcades under extremely hot conditions. Figure 8.17 indicates that all conditions clearly stood above the comfort zone but their horizontal alignment suggests that the operative temperature differential was kept to a minimum. Arcades, although slightly warmer than the adjoining streets, do not overheat. In fact, calculations have shown that there is no operative temperature differential between arcades and streets, while the arcade condition is, in terms of ambient temperature differential, 0.9°C warmer than the streets. The arcade considerably reduces air movement, which could unfortunately increase thermal discomfort. However, it lessens the dynamic discomfort created by the free wind in the open streets. Shops are slightly warmer than the adjoining arcade but again cannot be said to overheat significantly. The overall relative good performance of the arcade is probably owing to the shading effect of the roof and the shading provided by the nearby urban context.

Figure 8.18 illustrates rather strikingly the microclimatic advantages of the arcade type in winter. The two sets of clouds correspond to either warm winter days (dotted line) or cold winter days (solid line). On cold days, the great majority of the arcade conditions fall inside the thermal

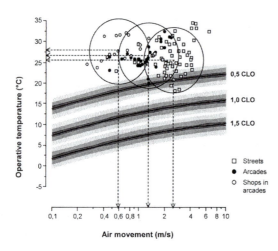

8.17
Summer comfort conditions for the entire arcade corpus

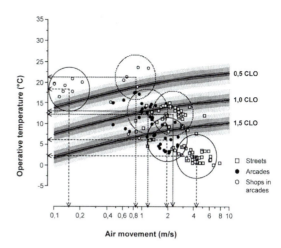

8.18

Winter comfort conditions for the entire arcade corpus

comfort zone, whereas the majority of street conditions fall well below it. An average operative temperature differential of 5°C gives the arcade a considerable advantage over the open streets. Moreover, air movement is reduced by 60%, which increases both thermal and dynamic comfort. The temperature differential is almost non-existent on warm days, which accords with the previous analysis for warmer ambient conditions. Interestingly, the thermal conditions of the shops are acceptable at all times with air movement being superior on warm days since tenants tend to keep the doors open.

In summary, the seasonal performance analysis has clarified the environmental advantages of intermediate environments such as arcades over totally enclosed or open urban spaces. It appears that the ambient temperature differentials between the street and the arcade increase steadily according to the severity of the season, from 0.9°C in summer to 1.6°C in autumn, and 2.5°C in winter while providing a constant two-thirds reduction from the open street air movement. The analysis emphasised the impact of major factors such as orientation, roof glazing, length and height to width proportion on the arcade environment.

Although the analysis has shown that the traditional arcade cannot ensure comfort in the summer, it does at least have the benefit of not overheating. This is of the utmost importance when considering the bioclimatic potential of such a spatial configuration as it challenges the common belief that arcades are unsuited to hot climates. Moreover, questionnaires and direct observations have identified a very practical behavioural adaptation to variable seasonal thermal conditions: *nomadism*. Tenants and patrons alike tend to occupy the basement (not shown on the drawings, but located under many of the shops' first floors) more often in summer as basement acts as a cool sink, and migrate to the first floor in the winter as hot air rises. Arcades can therefore be described as genuine bioclimatic urban spaces that encourage a dynamic interaction between the occupants and the variable thermal environment. Arcades favour a subliminal environmental adaptation that liberates the pedestrian from abrupt environmental changes that could lead to discomfort.

These results initiate a discussion on the extraordinary diversity of thermal transients, as experienced by pedestrians according to the urban form. Intermediate environments such as passages, *cours traversières* and arcades are genuine free-running transitional spaces that play a major role in the thermal scenario of the city, adding to the pleasurable diversity of the urban experience.

Return of the *flâneur*

Newly available technologies and the science of specification and prediction of physical ambiences are major forces in the current deterministic approach to the nature of built form and its use (Hawkes 1976). Yet, biological science suggests that environmental diversity, within a certain range of variation, is desirable to exercise the adaptive capacities of any living organism; a slight instability being the necessary condition to true stability (Potter 1971). A static approach to the notion of comfort is therefore not sufficient to account for the incessant successions of environmental stimuli of the real environment upon the individual.

There are numerous and straightforward means of creating environmental diversity. Among them, the segregation of environments by physical boundaries is the most obvious and probably the most efficient in terms of ease of control. There are, however, more subtle ways of achieving environmental diversity through spatial continuity that favour a progressive adaptation to a new environment. Sennett (1994) argues that the gradual desensitisation of the body by our increasingly isotropic environments has destroyed the awareness of the urban dweller to his/her physical and social milieu. In terms of physical ambiences, the current individualism or introspection of the urban dweller can be seen as the result of a combined lack of stimulation, excessive comfort in outdoor spaces, and excessive discomfort in outdoor spaces. Where too much movement in the street discourages sociability, too much comfort in the interior spaces also threatened sociability to a further extent. The ambiguous place of conviviality or the internal exterior, at the very centre of these two extremes, disappeared. In fact, very few spaces have been left in the contemporary city for the *flâneur*, who, not unlike a very rare and sensitive creature, needs a very specific environment to induce his incongruous behaviour.

The space of the *flânerie* took its most striking expression in the arcades of the late eighteenth and nineteenth century. In this pedestrian enclave, protected form the elements and the turmoil of the street, one slows down the pace and, suddenly, the level of awareness of people and space increases. Not unlike the basilica of the Roman baths, the space of the arcade is sumptuously ornamented to add to this phantasmagoria of having mysteriously, and without having even crossing a threshold, penetrated into a new realm. The arcade type appeared at the culmination of a slow evolution in the knowledge of the control of the environment by passive means. It also marked a transition in the mechanisation of architecture and the city.

Surveys of arcade environments have demonstrated that it is possible to achieve well performing passive urban environments within the variables of architecture alone. It appears to enhance our awareness of the environment, both physical and human, by the correct stimulation of our senses and a gradual bodily adaptation without impairing comfort. The arcade, albeit created in the eighteenth century to respond to such environmental problems as traffic congestion and noise pollution, still constitutes a valuable low-tech model for the creation of a more convivial twenty-first century city.

In an interesting portrait of Victorian society back in 1845, Disraeli wrote:

> the European talks of progress, because by an ingenuous application of some scientific acquirements, he has established a society which has mistaken comfort for civilisation.
>
> (Bradford 1983)

Disraeli suggests that progress, issuing from the great revolution of science and technology, had improved comfort but somehow lessened the real virtue of civilisation that is socialisation. Arcades bear the imprint of a pre-technological society that ranked its social values highly. There is no doubt that these extensions of the public domain into intermediate environments encouraged their social transactions and enhanced the overall experience of the city.

Acknowledgements

This chapter is drawn from a PhD dissertation, funded by the Commonwealth Scholarship Commission in the UK under the supervision of Professor Dean Hawkes at the University of Cambridge, and previously published work.

References

Baker, N. and Standeven, M. (1994) *The development of a personal sensor array for environmental monitoring in the PASCOOL comfort fields surveys*, Research report: Personal Condition Monitoring-PASCOOL Comfort Group, University of Cambridge.

Bradford, S. (1983) *Disraeli*, Briarcliff Manor, N.Y.: Stein & Day Publishers.

Flynn, J. E., Kremers, J. A., Segil, A. W. and Steffy, G. R. (1992) *Architectural Interior Systems: Lighting, Acoustics, Air-conditioning*, N.Y.: van Nostrand Reinhold.

Gibson, J. J. (1966) *The Senses Considered as Perceptual Systems*, London: George Allen & Unwin Ltd.

Hawkes, D. (1976) 'Types, Norms and Habit in Environmental Design' in March, L. (ed.) *The Architecture of Form*, Cambridge: Cambridge University Press.

Knudsen, H. N. and Fanger, P. O. (1990) 'The Impact of Temperature Step-Changes on Thermal Comfort', in *Proc. Indoor Air '90*, Toronto, Canada.

Lemoine, B. (1983) *Les Passages Couverts en France*, Alençon: Imprimerie Alençonnaise.

Penwarden, A. D. (1973) 'Acceptable wind speeds in towns', in *Building Science*, 8: 259–67.

Pignolet, F. (1994) 'Un code de calcul de prédiction du comportement thermique pour les espaces urbains', in *Proc. European Conference on Energy Performance and Indoor Climate in Buildings*, Lyon, France.

Potter, V. R. (1971) *Bioethics: Bridge to the future*, Englewood Cliffs, N.J.: Prentice-Hall.

Potvin, A. and Hawkes, D. (1994) 'Transitions in architecture: A bioclimatic approach based on thermal diversity', in *Proc. European Conference on Energy Performance and Indoor Climate in Buildings*, Lyon, France.

Potvin, A. (1996) 'Movement in the Architecture of the City: A study in environmental diversity', unpublished PhD thesis, University of Cambridge.

Sennett, R. (1994) *Flesh and Stone: The body and the city in Western civilisation*, London: Faber and Faber.

Chapter 9

The reverential acoustic

Tim Lewers

Introduction

We cannot see sound, and as a result the aural environment is not well understood. We are, however, constantly aware of our environment through all our senses and sound plays a significant role in conveying the nature of the space that surrounds us. Blind people, for example, can determine how far walls are from them by detecting the echoes from those walls. They can hear the presence of a lamppost as they approach it and they can detect open space by the absence of echoes. People with normal sight find that they, too, can do all of those things if they choose to develop their hearing through practice. For instance, musicians have highly developed hearing skills and when blindfolded have been known to be able to move around freely once they have the confidence to trust their ears. Before birth babies are subject to high noise levels from their mothers' bodies and after birth may suffer the trauma of relative quiet. Despite this noise, babies learn of the world outside the womb through sound and many mothers report that their babies react to sound whilst still in the womb and sometimes learn music well enough to respond to it after birth.

The human sense of hearing is a remarkable interface to the world around us. It is less easily deceived than the eyes, more sensitive than taste and touch, and, like smell, can evoke memories. Yet the manipulation of the aural environment is often limited to avoiding loud noise and providing warning signals. This is such a wasted opportunity. The aural environment is much more complicated than is usually acknowledged but it can be designed in just the same way as the spatial, thermal and visual environments. It is also subject to change and those changes play an important role in the way we perceive the world around us.

If acoustics is more important in the everyday environment than is usually believed, are there environments where the determination of the aural environment is recognised as primary to good design? One answer is that concert halls, discussed in more detail later, are regarded with particular reverence, and it is assumed that the acoustic design will be critical to the occupants' appreciation of their environment. Moreover, it is usually accepted that the spatial arrangement of a concert hall is dictated by the need to provide particularly favourable conditions for the performance and appreciation of music. This has not always been the case in western culture, as the concert hall is an eighteenth-century invention and we know that the history of music extends back though several millennia. The concentration today on the refined acoustic design of certain buildings and concurrent lack of attention to the acoustic environment elsewhere is certainly unfortunate. In this chapter the detrimental consequences for design of this separation and isolation of rooms with 'reverential acoustics' will be explored by examining the spatial, aural and visual sequences experienced on arrival at a range of concert halls but first the case will be made for the importance of diversity in everyday acoustic design.

Acoustic design

Acoustic design can usefully be split into three areas of concern: the background noise level, sound insulation between spaces, and the acoustic of the space surrounding the listener. In all cases there are ways of measuring the extent of the acoustic phenomenon involved and, as with all our senses, it is possible to have too little sensation as well as too much. The body can be too hot, or too cold, lighting levels can be too dim or too bright and so on. In addition, acoustics is an entirely subjective science and what is suitable for one person may be completely inappropriate for another, and therefore acoustic designers have to take a close interest in the way their particular audience responds to sound.

Background noise
A great many ways of measuring the background noise level have been proposed and each is appropriate for a particular task. There are many textbooks that describe the various units that are used, but we all have our own individual understanding of when a background noise is too loud or too quiet. Unfortunately, there is a very wide difference of opinion as to what constitutes an acceptable level of a particular type of noise. To some, the quiet of a country field, with no wind, far away from farm machinery and traffic, is idyllic, to others it is uncomfortable. To some, music is soothing, to others infuriating; teenagers find the latest hit single an acceptable background noise for doing homework, while their parents believe it makes

it impossible for them to concentrate. Someone who lives next to the sea and is subjected to a constant high noise level from the waves finds the quiet of the country disturbing. When on holiday, the city dweller is woken by the dawn chorus or the cock crowing, noises that are of no consequence to those used to the country, and yet the sound of traffic will prevent those used to country sounds from sleeping in the city. This author possesses a fine long case clock that chimes every half hour and whose constant ticking is a source of great comfort. Unfortunately, visitors do not appreciate its finer points and sometimes ask for it to be muted for the period of their stay.

Although sound is little more than a cyclical variation in air pressure, it carries with it information and the effect on the listener depends on the amount of information carried by the noise and its relevance, or lack of it, to the listener. An annoying noise may take the form of a rhythmic pattern, a melody, a tone, speech, a pattern of noises starting and stopping, a baby crying indicating a need, or any number of rattles, jingles and thuds that indicate that all is not well with something. Partly intelligible speech can be particularly annoying. The varying information component together with the effect of the variation of sound level with time and frequency makes the measurement of background noise level a surprisingly frustrating task. The background noise level of any noise can be measured in a multitude of different ways and yet the resultant numbers sometimes give almost no indication as to what is acceptable. The legal systems and standards authorities around the world have tried in vain to produce infallible guidance as to acceptable noise levels for various activities and yet there is no universally appropriate sound level and, indeed, no irrefutable way of measuring that level.

The one aspect of background noise control available to the designer is the use of masking noise. Noise can be used to create space and privacy, something that many people do naturally throughout the day. 'Walkman' type devices are used by people who have chosen to limit some of the excesses of urban life, using music to mask background noise. In that way they appear to create their own territory in which they feel more secure.

It is the deliberate use of background noise by designers that is of most interest to this discussion. Shopping malls invariably introduce background music to make the shops and arcades seem populated even when they are not. People buy more when they feel secure and an artificially high background noise, consisting of music, seems to be satisfactory to the majority. There are many malls that have a water feature in the central space. This is not only a visual focus but also a source of noise, and most importantly, a source of noise that carries very little information. The sound of water falling on water is comforting if it avoids sounding like a dripping tap and more like a waterfall and the effect on the space is to make the area around the shopper appear safe and the environment acceptable. Switch the fountain off and people feel less secure, the mall is less inviting and shoppers keep to the walls, stay a shorter time and buy less.

Moving water is used in many environments, one notable example being the approach to the Louvre museum in Paris. The glass pyramid that forms the entrance to the museum dominates the central courtyard and is surrounded by water features. The resultant noise increases as one approaches the pyramid and the feeling is of being transported to 'Louvre Space' in the sense that it is a different space from the centre of Paris. By the time visitors have arrived at the entrance they have walked for several metres through a zone whose background noise level is dominated by the sound of the waterfalls and not by the traffic which is only a few metres farther away. The effect is to calm the visitors before they enter the museum and to smooth the route from city to museum (Figure 9.1).

The most common use of the technique of artificially increasing the background noise level is found in open plan offices. Offices without walls are spatially efficient and promote communication and therefore productivity. However, the major drawback for their occupants is the lack of acoustic privacy. Colleagues need to be able to communicate when they are a few metres away from one another, but they should not be able to hear intelligible speech from workstations farther away. People talking when present in the office is slightly irritating, but hearing only one side of a telephone conversation is far worse, and overhearing partially intelligible speech is not conducive to concentration.

Open plan office design has a well developed acoustic strategy for success that is often neglected or ignored. The office should have a thin carpet, an acoustically absorbent ceiling, acoustically absorbent screens surrounding each workstation and a relatively high background noise level. It is this last requirement that is often ignored because it is counterintuitive.

9.1
The Louvre main courtyard and its Pyramid entrance, Paris

Noise conditioning systems, or the production of noise via loudspeakers, provide the required background noise if the sound from air-conditioning systems is insufficient. If well designed, these provide a bland but usefully high background noise level that does not disturb and creates privacy. This technique is little understood and often wrongly omitted from the design of offices with detrimental effects on the acceptability of the office environment and therefore on staff turnover and productivity.

Sound insulation

As with background noise level, achieving the correct degree of sound insulation is a vital part of the quest for a suitable aural environment. While it may be true that the greater the sound insulation, the greater the cost of the construction, if too little effort is invested in either the design or construction process, too low a level of sound insulation can be very difficult to rectify.

There are occasions where too much sound insulation is a client requirement when lower criteria would produce a more pleasant, more diverse and less sterile environment. This has frequently been the case in music schools. Of course, music practice and teaching rooms should not disturb each other, but there is no reason why the circulation spaces need be very quiet. If the doors are not over-specified they can transmit more sound to the benefit of the life of the building. People moving round the corridors can then anticipate the life and activity in the teaching spaces. Similarly, drama schools should surely reveal something of their activities in this way, rather than portray a lifeless, insipid environment.

Offices in the city often have no openable windows, shutting out the very life that makes the city such a vibrant place to live in. The inability of office workers to adjust their connection with the outside world is a huge frustration for many. Air-conditioning systems often preclude this option but designers should remember that occupants might want a little noise.

There are, of course, many more examples of too little sound insulation. Hotels are often under-specified and the sound of the occupants of an adjoining bedroom is frequently disturbing. Individual offices rarely seem to have adequate sound insulation between them, usually because little thought has been given to the passage of sound around a partition, especially through the ceiling void. They then suffer from a lack of privacy and private conversations are overheard. Privacy is closely related to the combination of background noise level and sound insulation and the absence of adequate sound insulation can sometimes be compensated for by an increase in noise level.

Whereas modern building regulations require builders to achieve an appropriate degree of sound insulation between adjacent dwellings in both 'new-build' and refurbishment projects, the achievement of adequate standards of sound insulation in existing buildings remains a daunting challenge. The near impossibility of dramatic improvements to the sound insulation between existing dwellings can frequently be very frustrating to occupants and advisers alike. Sound insulation is not very well understood by the construction industry and even less well by clients. A more complete understanding of the principles involved, which are simple, could benefit the design of many ordinary buildings but it is likely that over-specification will remain a problem in some new buildings as a result of clients' anxiety about sound isolation.

Room acoustics

The acoustic of a room is one of the critical components of its environment. Yet the acoustic of a multitude of ordinary rooms is unnecessarily neglected and the acoustic environment could be much improved if some simple alterations were made where necessary. Ordinary rooms should have an appropriate reverberation time. That is to say, a room should allow the sound to persist for the length of time that the average occupant thinks it should. We have all experienced rooms that seem unduly dead or live and these can be unsettling. This is therefore another acoustic criterion that has boundaries at both ends of the scale. The reverberation time is nevertheless just one of the features of the acoustic of a small room and not always the most important. The presence of flutter echoes (sound bouncing between parallel walls) is often mistaken for excessive reverberation. Such echoes are disturbing but easily eradicated by placing absorbent material on one or both walls. Many domestic rooms, offices, consulting rooms, meeting rooms, interview rooms and other small cellular rooms, show signs of flutter echoes, the removal of which would vastly improve the experience of the room.

Some rooms show signs of room resonances and certain tones predominate, which can make music unpleasant to listen to, whether live or recorded. Again, a modicum of absorption in the right place rectifies the problem. Individual echoes can also occasionally be heard in rooms, especially if they have concave surfaces. Occasionally, diffusing the offending reflections will eliminate annoying echoes and resonances. Adding diffusion to a wall can be achieved by adding furniture, bookcases, works of art or indeed any non-flat object. The trend to minimal modes of living has led to some dreadful domestic acoustics when all that is required is domestic clutter and rougher walls.

Many other ordinary rooms would benefit from some simple acoustic design. Restaurants could benefit from absorbing ceilings, but many have hard and high ceilings which cause the background noise level to creep higher and higher as people try to make themselves understood to their fellow diners. Sports hall designers often ignore the acoustics and then wonder why people have to be very close to one another to be understood. Museums and art galleries often have too live an acoustic, which makes them intimidating to newcomers. Dining halls would all benefit from absorption on the ceiling, as would rooms for cocktail parties and receptions.

Just a little care and a small amount of knowledge would improve the acoustic of many rooms enormously.

'Reverential' acoustics

The human race has been attracted to the acoustic of the cave for perhaps hundreds of thousands of years and much has been made about the acoustic phenomena that can be heard in caves. Indeed, take children into a reverberant acoustic, such as a simple railway arch, and they will invariably make a noise and listen to the effect with delight. As soon as buildings could be built, imitation caves were constructed, normally for religious purposes, and we continue to recognise that the acoustic of the cave has a culture of significance and mystery.

The power of the state can be symbolised by the reverberance of the foyers to parliaments, law courts, palaces and town halls. Churches, temples, mosques and synagogues continue to be built with a reverberant acoustic despite the problems with speech intelligibility that result. While reverberant railway stations announce to travellers that trains are important, shopping malls and arcades are deliberately reverberant to enable sounds to persist for long enough to increase the background noise and create privacy. National art galleries and museums seem always to be designed with reverberant galleries as well as foyers, despite the fact that this often results in intimidating buildings that discourage visitors.

Reverberance continues to have mystery in today's culture. In films and television, if the director wants to portray an unnatural or unsettled environment, it is easy to add reverberance to the soundtrack and immediately the viewer is put on edge.

Western classical music has evolved in the acoustic of the cave and it is one of the few forms of culture that requires a reverberant acoustic. Pop, folk, jazz and theatre all require the acoustic of the open air — the dead acoustic associated with small intimate rooms. People feel comfortable in such rooms, the lights can be dimmed without causing concern and the mind can be led by the entertainment with a degree of ease. Not so in the modern day, purpose-designed concert hall. In visual and acoustic terms these buildings seem to portray the music as difficult and intimidating entertainment for the elite by providing space whose appearance and acoustic are more likely to encourage people to feel as if they have entered an artistic world apart than to relax in enjoyment of the performance.

The concert hall

Classical music evolved in the houses and palaces of those who could afford to employ musicians and composers. Later, that music was played in tennis courts and drill halls, these being the only ones available to the general public that resembled the great halls of the aristocracy. Later still, in the nineteenth century the invention of the concert hall allowed the specific needs of performance to a wider audience to be addressed. It was only at the beginning of the twentieth century that the science of acoustics was established.

Auditorium acoustics is a science that has contributed much to the design of modern day concert halls and it is now possible to design a hall with a suitable acoustic for orchestral music for up to 2200 people with confidence. Mathematical, computer and physical modelling techniques have been devised that enable the designer to proceed with a reasonable belief that the first night will not be a disaster and that the hall will be enjoyed by audiences for the foreseeable future.

Audiences are, however, smaller now than they once were and the image of classical music remains that of a difficult, elitist and remote form of entertainment. It is perhaps strange, therefore, that despite the existence of concert halls whose dimensions and detailing have been precisely determined by science to produce an ideal acoustic environment, audiences will go to hear all kinds of music in theatres, churches and cathedrals with inappropriate acoustics and hard seats, or even in the open air.

Some years ago the author was involved with a programme to visit many of the major concert halls in the UK and collect data about their subjective qualities. This data was then correlated to objective measurements indicating how sound behaved in these halls and some useful results emerged. The listeners were all acousticians in order to ensure an educated group that knew the terminology of the data collection process and who were interested in visiting the halls. One evening during the concert interval, a professor of acoustics pointed out that we had all travelled with ease to the venue, parked the car and walked to the entrance without trouble, had a pleasant cup of coffee before the concert and a decent glass of wine in the interval. We had found our seats with ease and then listened to a famous international cellist play magnificently for two hours. We were then asked to fill in a form about the acoustics of the hall. His response was that acoustics was nonsense.

This is a fascinating comment from such an eminent man because it puts the science of acoustics in perspective and perhaps opens a new way of thinking about 'reverential' acoustics that has been missing since the invention of the concert hall.

Three London halls

The Royal Albert Hall (Figure 9.2) is in west London, between Kensington Gardens to the north and a site that was developed after the Great Exhibition of 1851 to the south. This important venue opened in 1871 and has had a famously dreadful acoustic from the outset, due to its size (it seats 5222) and

9.2

The Royal Albert Hall, London

its oval shape. Several major remedial measures have been applied since its opening and these seem to have mollified the worst acoustic problems. Each summer the hall is host to the Proms, a two-month long festival of music during which the hall is frequently filled to capacity. The promenaders stand, often for several hours, and listen to the concerts with a degree of concentration that is sadly missing in other venues. Even though the acoustic remains poor by today's standards, the building itself, with its exuberant red and gold decoration and unusual seating plan (all members of the audience are aware of one another throughout the performance due to the oval seating arrangement) is London's greatest example of High Victorian concert hall design.

The Royal Festival Hall (Figure 9.3) was opened in 1951 and was the first major cultural building to be constructed in Europe after the war of 1939–45. It stands on the south bank of the river Thames and enjoys a fine view of the river and the buildings that line the north bank. It was built to the highest standards of the time using modern materials. The latest scientific advances in acoustics were applied, and everything was carried out in a spirit that epitomised England's cultural rebirth after the war. It remains a fine building and is recognised as a major landmark in the history of architectural design. However, the acoustics have been criticised since its

9.3
The Royal Festival Hall, London

opening because the acoustic of the space is too dead and has too much clarity. It is noteworthy that the audience all faces the stage and the colours are far removed from Victorian extravagance. Audiences for the Festival Hall have been poor in recent years and a proposal has now been tabled to refurbish the hall with the aim of attracting back the audiences that it once enjoyed.

The Barbican is a concert hall buried within the mainly residential complex also known as the Barbican, just north of the City of London (Figure 9.4). The development was begun in the 1950s but the hall did not open until 1982. From the outset it has been criticised for its poor acoustics, but a recent refurbishment has led to better reviews. It is a wide hall with the audience in two main groups of seats, one above the other but without a

9.4
The Barbican Concert Hall, London

balcony overhang. The seats are comfortable and spacious and the acoustic, though not flatteringly reverberant, is pleasant enough. The Barbican is the home of one of the finest orchestras in the UK and audiences for these concerts are good, but it is not a hall that has attracted a great following in the manner of the Royal Albert Hall.

In cultural terms, as much as in terms of mood or ambience, the contrast between the three halls is great. None of them features in anybody's top 10, or even top 40, of the finest halls in the world. None of them would be the optimum shape for a concert hall if it was to be designed today, and only the Royal Albert Hall is well known outside the world of classical music. Might closer consideration of the aspects of the concert-going experience that fuelled the original insight that acoustics are nonsense throw some light on this issue?

The Barbican does not have a facade that is visible to a passing pedestrian, indeed concert goers have to be encouraged to find the entrance from the public transport stations by yellow lines drawn onto the walkways. The entrances are dark and low, the foyers confusing and cluttered. The route from the foyer to the hall is not clear and finding one's seat can be traumatic for the novice.

The Royal Festival Hall is always approached indirectly, even from across the river. The normal method of arrival is to its rear across bleak paving and busy roads. The hall can be seen from some distance but as you get close to it, it disappears only to reappear when you are right up against it as a series of laconic blank walls. The entrance has lost its coherence due to the introduction of raised walkways but arriving in the foyers leaves the visitor confused because, although the hall is above the foyers, the stairs are hidden to the sides.

In contrast, the route to the Royal Albert Hall is easy to find, and arranged as a generous celebratory sequence of exterior and interior spaces. Once entitled 'the village hall of the empire', the hall has a uniquely privileged cultural status amongst audiences because of the range of performances for which it has provided the setting. Whether staging moments of national importance or of more personal delight, it has provided and still provides a magnificent arena for musical, sporting, political and theatrical events. Despite the problematic acoustics of its main auditorium, it is a building that generates great excitement. Take any child to it and they sense the atmosphere of *joie de vivre* and excited expectation as they approach. On arrival, its dome can first be seen between the roofs of the surrounding buildings, but it does not spring into view, rather it gradually appears, becoming in the process an obvious focus for orientation through the nearby streets. Once inside the building, restricted glimpses into the auditorium from the surrounding foyers clarify the location of the hall and build the tension. In a splendid theatrical gesture worthy of the finale to this sequence, the main entrances leading on into the space then open up the greater scale of the hall in one dramatic moment. To quote the, great American acoustician Leo Beranek:

> the orchestra played Tchaikovsky's Overture 1812. The hall demonstrated its well-told ability to stir the emotions of its patrons ... The echo enhanced the gunfire, the chimes and the fanfare ... Above all, the great organ sounded like the voice of Jupiter. The audience was left breathless and tingling. It is for these moments of ecstasy that the Albert Hall continues to exist.
>
> (Beranek 1996: 308)

No hall epitomises that acoustics is nonsense more than the Royal Albert, for in a larger sense, no hall can better contain and enhance a great performance of the larger repertoire than this Victorian masterpiece.

Two opera houses

While the acoustics of any auditorium will always be of fundamental importance, it is clear that allowing acoustic considerations to dominate design discussions will not, in itself, guarantee a satisfactory performance space. Discounting the significance of how a performance space is encountered experientially ignores the nature of the events for which the building provides a setting. A comparison of the very different entry sequences into the Paris and Sydney Opera Houses highlights the environmental issues at stake as far as this issue is concerned.

Extravagant in form and scale, the Paris Opera House provides a fittingly dramatic finale to the Boulevard Opéra. On arrival the visitor is in no doubt about the location of the entrance or the principal route to the main performance space. That route is ordered by way of a succession of openings and veils that encourages the audience to move upwards through an arrangement of stairs and archways that deliberately heightens a sense of excited anticipation (Figure 9.5). By virtue of this arrangement, the view into the auditorium, an enormous room of great splendour which demands a great performance, is withheld until the last possible moment. It seems to matter little that some of the seats have a restricted view of the stage and there are a few that have no view at all.

The Sydney Opera House is a major cultural icon of the southern hemisphere and undoubtedly the most famous building in Australia. Standing majestically on a peninsular that juts into the harbour, its complex shell-like form attracts visitors from all over the world (Figure 9.6). After following a road from the city that leads directly to the building, climbing the stairs to the entrance is for many the end of a pilgrimage. It is sad that visitors are faced

9.5
The Paris Opera House

with an internal spatial sequence whose stark uninviting character fails to fulfil the promise of the building's strikingly exuberant exterior.

Neither of these two buildings has a great tradition of opera like that of La Scala in Milan or The Metropolitan Opera in New York, and yet, as examples of urban design whose intention is to attract and welcome an audience they are fine examples. Their lack of success in housing opera is largely due to the technical requirements of a modern opera house and not to faults in their design or indeed acoustic. But they do show that the journey to an artistic event can be as good an experience as the event itself.

9.6
Sydney Opera House

Conclusion

This chapter has highlighted two major issues: on the one hand, the importance to architectural design of consideration of ways in which sound can create or subtly qualify the experience of buildings, and on the other, the importance of the overall context and the disposition of appropriately paced experiences in and around buildings where musical performance takes place. The difficulties that have arisen as a result of an almost exclusive focus on acoustic performance criteria in discrete spaces like auditoria point to the potential impoverishment of design if environmental criteria are not intelligently interpreted and communicated. Currently, spatial design is almost always discussed in exclusively visual terms. This needs to change. Closer interest in how sound can be veiled or reinforced by material means and how space is defined and communicated through sound should allow designers to better describe aural diversity and to address how it can be achieved architecturally. In addition, it is clear that architects and acousticians should be looking to develop a better understanding of how people hear, make sense of, and judge the aural environment, both at rest and in movement, i.e. from a sequential as well as a static experiential viewpoint. It is to be hoped that in the future, the idea that an architect is redefining the 'soundscape' as much as the 'landscape' of a site, the 'unbuilt', as much as the 'built' landscape, will prompt a much richer, more nuanced conversation about what aural diversity is possible or appropriate in a given situation and how it may be coordinated with other design intentions.

Reference

Beranek, L. (1996) *Concert and Opera Hall. How They Sound.* Acoustical Society of America, American Institute of Physics.

PART 4

Interior

Chapter 10

Environmental diversity and natural lighting strategies

Mary Ann Steane

Introduction

Light in spatial and temporal terms — whether in relation to a sequence of spaces, movement through them, or material qualities — is the visual mechanism that creates experiential diversity in architecture. The emphasis of guidance and regulations has tended to be on providing illuminance uniformity, yet many spaces do not need such strict control, positively benefiting from the stimulation induced by variety in the quantity and quality of light. Natural light is, however, neither a constant nor necessarily a constantly varying resource. In the search for diversity its quality and dynamic variability, dependent on latitude and local climatic factors, need to be taken into account. This chapter describes by way of some well-known building case studies a series of considerations that demonstrate the relevance and richness, both climatic and experiential, of strategies promoting lighting diversity in architecture. These strategies range from those that respond to patterns of movement (spatial and solar), through those that manipulate the transmission of light, to those that exploit differences in the specularity of surface finishes. A specific focus on the use of materials and their visual characteristics in Terragni's Casa del Fascio in Como provides a clear demonstration of the specific and conscious role played by luminous diversity in shaping our experience of architecture.

As this discussion will demonstrate, the pursuit of visual diversity requires designers to focus on how natural light may be used to define or qualify spatial sequences and to take a close interest in how buildings reconfigure natural light over time. While it is true that variety in lighting conditions is the inevitable consequence of any spatial sequence, the issue of how and why such sequences in light may be deliberately shaped deserves attention in its own right. Here the lessons provided by designs whose conscious aim has been to orchestrate visual diversity will be examined. A range of specific examples will be used to illustrate how it is possible to coordinate lighting design strategies with other spatial, programmatic and environmental ambitions. The attempt to establish ends will inform a discussion of the means used to achieve luminous diversity.

At the outset a few words are necessary on the design implications of this issue. Diversity in the visual environment is experienced as contrast in the level of brightness of different areas of the visual field. A visual environment may thus have lower or higher spatial diversity, i.e. it may be more or less evenly bright but it may also have higher or lower temporal diversity, i.e. its distribution of light, and thus of brightness contrasts, may fluctuate more or less over time. Spatial diversity within a room is primarily affected by the basic geometry and colour of surfaces, as well as by the sizing, location and glazing of apertures. Thus large, white, rooflit spaces with diffuse lighting will tend to have lower spatial diversity (diffuse light from above is scattered more evenly internally by high-reflectance, white surfaces) whereas low, deep, darker, directly side-lit spaces will have higher spatial diversity (direct light from one side is scattered less evenly by low-reflectance, dark surfaces). Though the visual interest of individual spaces may on occasion be of significance, the pursuit of spatial diversity is likely to have greater significance for building-scale design strategy. Temporal diversity on the other hand, is primarily a room-scale issue dictated by the degree to which movement of the sun (and on occasion the wind) is able to animate the lighting conditions within a space. It is therefore the orientation, geometry, and location of apertures with respect to room surfaces that influences whether temporal diversity is high or low over particular periods of the day or year.

What is perhaps less well appreciated is the extent to which the visual environment may be more subtly transformed through close attention to the light transmission qualities and surface textures of materials. When in sunlight, translucent materials appear to glow and shadows may be projected through them while the light they transmit is more diffuse. When seen from particular viewpoints in appropriate conditions, shiny or glossy materials gleam or glisten with reflected light. Where these effects are exploited, the entire appearance of a space may be transformed as sunlight strikes a particular material or as the viewer moves from one location to another. As will be examined in detail later, the intelligent deployment of such material qualities can thus increase visual diversity by creating spaces whose 'visual stability' is in question.

The discussion which follows looks at a series of case studies where visual diversity has been achieved in a range of ways for a range of purposes. It has been organised according to the series of sub-headings that arise out of this analysis of the way in which temporal and spatial diversity may be experienced and achieved:

1   lighting induced movement
2   solar movement
3   the specularity of surface finishes.

The exploitation of spatial diversity: lighting induced movement

Martin and Owers at Kettle's Yard: a slow unfurling of space in light
Originally conceived as part house and part gallery, Kettle's Yard was formed in 1957 by art collectors Jim and Helen Ede. By knocking four derelict Cambridge cottages into one, they set out to create a series of domestic settings for the display of their unusual collection of art works and found objects. In 1970, a sympathetic transformation and enlargement of the original house to provide further gallery space was carried out by Leslie Martin and David Owers. What is most remarkable about the final arrangement is the orchestration in light of the spatial composition that unites house and gallery. The relatively dark cottage entry hall leads initially to a pair of small but well-lit, bay-windowed ground floor rooms, and then via light from the turn of a spiral stair up to a larger first floor room that commands longer views out over the neighbouring common. A brief visit to a more constricted space in the attic then prefaces a slow descent through a series of rooms that gradually increase in size and degree of enclosure. The light is increasingly more diffuse, the mood increasingly more quiet and unhurried. This slow unfurling and opening up of the interior offers a nautilus-like spatial arrangement in which a series of chambers grow in size as one moves outward from the centre. Along this route a gradual rise in light levels to the bright crescendo of a first floor rooflit gallery is followed by a gradual falling away into the half light of the darker regions of the ground floor (Figures 10.1–10.4).

A range of spatial settings and lighting conditions are thus provided for the many objects and paintings of the collection. Their informal but precise placement ensures that the art on display is always very much to hand, and not at all removed or remote, while the visitor is constantly invited to enjoy the way that the location of an object or painting exploits and qualifies light and shadow. Sensitivity to the idea that light, objects and space mutually define one another is also apparent in the choice of surface finish for the walls of the new extension. A white finely textured plaster surface means that in spatial terms the walls are simultaneously background and foreground, materially remote yet tactile. The large flat surfaces recede into neutral whiteness behind the paintings and emerge in a gauze of shadow that dramatises the fall of light from above.

10.1 (above left)
Kettle's Yard house ground floor

10.2 (above middle)
Kettle's Yard house first floor

10.3 (above right)
Kettle's Yard gallery extension first floor

10.4 (left)
Kettle's Yard gallery extension ground floor

MeadowcroftDernie: the act of seeing

Architecturally, the design of MeadowcroftDernie's Clerkenwell gallery provides a dramatic visual sequence between the street and the heart of the block: beyond a narrow, shaded facade the visitor crosses a subdued window-lit upper gallery towards a stair and a well of light that connect the basement to the sky (Figure 10.5). At basement level a door leads on into a much darker but dramatically lit inner sanctum (Figure 10.6): the blue rooflit dining room of the gallery's owner. (The rooflight area here is much smaller.) The architect's juxtaposition of a tall, strongly lit, white painted gallery space with a tall, much darker, inner room is only experienced occasionally by

10.5
Clerkenwell gallery

10.6
**View into dining room from
basement of gallery**

friends of the client and guests at exhibition opening parties. Just as on entering a cave, everyone who arrives from the gallery has to stop and wait for their eyes to adapt to the much dimmer light of the dining room before they can 'see' where they are. The decision to ensure a pause in movement, and an adjustment of vision at this point in a spatial sequence where works of art are the object of attention is unexpected yet telling. It reflects the architect's intelligent appraisal that by asking the viewer to undergo a significant degree of visual adaptation, the very process of seeing is made the focus of experience.

*Asplund at the Woodland Cemetery: final journey through
a landscape*
A typical visit in winter to Asplund's 'Woodland Chapel' involves an extended walk through a snow-covered landscape in which movement, views and light have been carefully coordinated to encourage a quietly reflective state of mind.

Beyond the dark stone-walled entrance to the cemetery, the rising sweep of the white hillside leads the eye past a line of smaller chapel buildings and low trees to the tall wooden cross that stands at the summit. Just beyond and to the left of this cross, the austere columnar frame of a much larger building prefaces the entrance to the main cemetery chapel. In this initially more open cemetery landscape it is the even white covering of snow that ensures that the fall of the ground is simply outlined against the darker snow-laden trees and that everywhere the ground is very bright. Typically, only the slowly rolling grey clouds provide visual animation. After climbing the hill, some relief from the bright surfaces of snow is given by the large portico. From without it defines an area of deep shadow pierced by light,

10.7 (above left)
Woodland Cemetery entrance

10.8 (above top)
View down hill towards entrance

10.9 (above)
Portico to main cemetery chapel

10.10
View from Woodland Chapel entrance back to gate

from within its columns slice the landscape into a series of bright abstract tableaux that cut the view down to size and give it a more human scale. From here the journey continues down a wide track on the other side of the hill between densely planted conifers. Views are more restricted, light levels are noticeably lower and the darkness of the space under and between the trees becomes more apparent. After a few minutes' walk, a pedimented gate in a low wall quietly announces the presence of the Woodland Chapel among the trees to the left of the track (Figures 10.7–10.10).

The simple pitched roof building is only fully visible once one has traversed the much darker path under the trees and emerged into a small break in the wood. A relatively low and therefore relatively dark, white-painted columnar entrance area allows mourners to gather and look back before they enter through the heavy wooden doors and wrought iron screens of the deep threshold. Within, darkness prevails but for the diffuse light coming from the pale surface of the dome in the ceiling and the bright points of warmer flickering candlelight that punctuate walls which are otherwise completely in shadow. A circular abstract sky, the location and materiality of whose surface is in question, circumscribes the mourners who gather around the coffin in the much darker space below. Gradually, but without difficulty as a result of the careful orchestration of light conditions along the journey, visual adaptation to the very low light levels occurs and a spatial stillness is experienced which seems wholly appropriate to the character of the gathering.

High temporal diversity: solar movement

Herzog and De Meuron, Dominus Winery, California: filtered sunlight, fractured walls

The first floor corridor-galleries between the administration area and the main fermentation hall of Herzog and De Meuron's Dominus Winery are contained between an outer wall construction composed of a rainscreen of gabions (galvanised steel wire cages containing basalt stones) outside a glazed steel frame, and a lightweight inner screen lined with wire mesh. The use of this layered construction has an extraordinary result for the interior: a heavy building is fractured by light, stone walls are lit up from within. Strong sunlight is projected through the interstices between the stones across the relatively dark corridor gallery, either onto wire mesh screens, or far below onto the concrete floor of the main entry hall. It is a striking effect for being so unexpected, a rough deeply textured wall of the type one might normally associate with a civil engineering project has a craggy beauty from the outside and reconfigures through light the otherwise simple space within. As the sun moves around the building the pattern and character of the sun patches and the appearance of the walls keeps transforming through the day. Like sunlight coming through foliage, the size and degree of definition (the penumbra) of each sun patch constantly varies with the angle at which the sunlight strikes and inter-reflects within the wall.

The construction strategy is an intelligent and poetic solution to the problem of coordinating an integrated response to the various environmental requirements of the building in the context of a climate that is hot by day, cold by night and sunny for much of the year. The three sizes of mesh and three sizes of stone which vary the density and transparency of the enclosure not only give the building an intriguing diversity of lighting conditions, but help to provide parts of it with an appropriately high level of thermal mass. (Although the offices are air-conditioned, the rest of the winery building is naturally cooled and ventilated.) This allows the performance of the envelope to be adjusted to the needs of the various processes going on within. While they appear very similar from the outside, other sections of the walls are heavier and more impenetrable to light and heat than those beside the administration area. Where wine storage takes place, for example, much smaller stones are used in finer mesh cages to produce cool, completely dark, naturally ventilated spaces.

Gabriel Poole's Tent House, Eumondi, Queensland: glowing surfaces, projected shadows

In his Tent House at Eumondi in Queensland, Poole embraces a new attitude to the light and landscape in which the building is located. Wishing to construct a space out of the projected light and shadow of the bush, he has designed a lightweight steel structure that supports a raised house whose roof, and most of whose walls, are canvas. Sitting high up in the tree branches, it exploits the prevailing breezes that are so welcome in a tropical climate and opens itself to the view. The canvas walls may be rolled up and translucent vinyl doors may be slid back over the exterior fixed walls. Because the canvas walls and the twin canvas membrane of the double roof are translucent, the house is flooded with light during the day. The relatively low contrast between interior and exterior light levels that this use of canvas engenders makes viewing the bright landscape outside much more comfortable than is possible from within a more solid opaque enclosure. As Walker points out (Walker, 1998), throughout the day wind and sun constantly modify the pattern of light and shadow that plays across the canvas surfaces, defining an ever-changing interior whose boundaries are constantly in question.

Lighting diversity and the specularity of surface finishes: animation through visual instability

Guiseppe Terragni's Casa del Fascio, Como: gleaming surfaces, reflected images

Considerable photographic and documentary evidence exists to confirm that Terragni thought very carefully about how the surface finishes of the Casa del Fascio in Como would affect the character of the visual environment (Terragni 1936a). Their effect on the lighting was a matter of deliberate, if intuitive, design and not a side issue or an unforeseen consequence. It is for this reason that the building usefully illustrates how the optical properties of material surfaces may be exploited in daylighting design.

Commissioned by the local fascist party as its new headquarters, the Casa del Fascio was completed in 1936. A new building type that provided both a meeting place for a political party and most of the offices needed to organise and run local government, no explicit precedent existed for its arrangement in plan and section or for its lighting environment. Terragni's response was to design a 'palace' of offices, whose entrance foyer would act as a key public space in the town. Organised on four floors around a central double-height space at ground level and a second floor open courtyard, the building's public entrance facade faces south-west onto a square paved with white marble. It has street facades facing south-east and north-west, while at the rear its north-east facade looks onto an adjacent building across a narrow service yard (Figures 10.11 and 10.12).

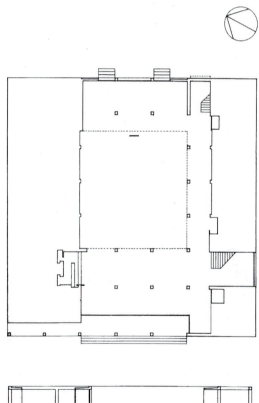

10.11
Ground plan of the Casa del Fascio, Como

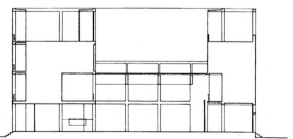

10.12
Section of the Casa del Fascio, Como

In scale and character, the building's key internal space lies somewhere between a small covered courtyard and a large rooflit internal hall. Light enters directly through three main apertures: the glass wall of entrance doors spanning the width of the hall at the front, the horizontal line of windows at the rear and the narrow central rooflight that runs from front to back of the hall. Some additional light is borrowed from two well-lit staircases at the rear and the south-east street facade. Further diffuse light is provided through the rest of the hall ceiling. The glass block panels that make up the stepped ceiling section glow and shimmer when struck by sunlight. Below them, the three deep structural beams that straddle the hall baffle the bright light source of the slot window, providing partial reveals that mitigate the contrast in brightness between the transparent glass and the glass block panels. Some light also enters the hall via glass-lens fanlights above all the doors, and glass-lens windows at the top of the first floor office walls (Figure 10.13).

According to Terragni, the hall to such a building needed to accommodate public meetings and it is for this reason that it addresses the main public arena of the adjacent square. The space is thus oriented to a view looking south-west, towards Como's cathedral and the dense medieval fabric of the city beyond. On entry one passes from the open square up a small flight of steps, under and through an external loggia to the main entrance doors that lead to the reception office. Once inside, the windows of the building's most important meeting room are visible at first floor level on the far side of the hall. To reach this room following the formal entrance sequence one climbs the main stair beside reception. It is the precise specification and disposition of the surfaces on this route in relation to those of the hall as a whole that illustrates Terragni's careful approach. For Terragni the specularity of the materials mattered. Sensitive to the need to coordinate spatial and material decisions, he

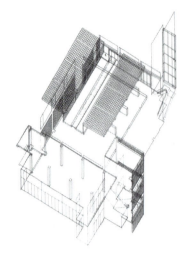

10.13
Axonometric diagram of the main hall of the Casa del Fascio indicating location of daylight sources

10.14
Ground plan published by Terragni in *Quadrante* **indicating key routes through building (***Quadrante***, 1936, 35(36): 55)**

10.15
Aqueous light: the view towards the rear of the hall at first floor level

understood how to key the geometry and level of specularity of different surfaces to the location of light sources and to the route of the viewer on moving into or out of the building (Figures 10.14 and 10.15).

TERRAGNI'S REPORT: ENVIRONMENTAL DESIGN PARAMETERS

The October 1936 issue of the Italian architectural journal Quadrante was entirely devoted to the documentation of the recently completed building (Terragni 1936a and b). Photographs, drawings, diagrams and lengthy analysis allowed Terragni and his colleagues to indicate the range of issues and ambitions which had driven the design. Dispersed through the text are indications of Terragni's environmental concerns, amongst which the question of how natural lighting was to be provided was very prominent. For example, a series of famous diagrams based on those of Neufert (1936) describe solar penetration through the seasons at each of the four facades alongside a graph in which the hourly variation of air temperature for the solstices and the equinoxes is indicated (Terragni 1936b: 43). Interception of the hot afternoon summer sun thus justifies the use of the loggia on the front facade. The visual environment of the hall is discussed with reference to the late decision to include the long slot window at the centre of the soffit:

> Note the longitudinal sheet of crystal of enormous thickness (33 mm) that allows a view of the landscape and the sky, so that the soffit in glass-block does not weigh down such a vast room with the diaphanous uniform light of its reticulated surface.
>
> (Terragni 1936a: 24).

Even the myriad reflections of the foyer are referred to obliquely while the numerous annotated photographs provide sufficient evidence in themselves that Terragni cared about what he called 'the remarkable effects of specularity' (Terragni 1936a: 16). Indeed, one of the most curious images in the entire issue is not of the space but of its reflection in a black wall surface. Others describe the doubling effect of key mirroring surfaces such as that of the building's most important piece of furniture, an enormous glass meeting table, and the walls of the main stair (Figure 10.16).

The use of marble in the entrance area receives the following comment:

> The stairs, the floors of the hall and of the entrance area, the pillars and the soffit of the entrance area … boast large marble surfaces of a colour in accordance with the use of glass and metals, that multiply with the specularity and the spare lines of their natural veining, the sense of that 'pure' beauty that only valuable materials can provide.
>
> (Terragni 1936b: 51)

10.16
Reflections on the wall of the main stair

It is important to point out that in three key respects the lighting environment encountered currently in the Casa del Fascio is different to that described in Terragni's technical report. First, the glass-block ceiling has been covered with panels of corrugated glass that have reduced the amount of light entering the main hall through the roof. Second, the colours of the interior paint finishes have been changed, possibly in response to the lower availability of light. Originally the plaster surfaces of the interior were painted a sea-green colour where they are now a bright white. The gloss painted doors and radiators were once black in colour where all the doors and most of the radiators are now grey. Third, the surface of the black marble ceiling has been badly damaged by fire. This has significantly reduced its degree of specularity.

SURFACE TEXTURE AND LIGHT

Before examining the consequences of these decisions for the lighting strategy, some comments need to be made on the variation in visual character of different surface finishes. Depending on their surface texture, materials reflect light more specularly or more diffusely. If they reflect most incident direct light precisely in one direction, reflected images may be seen in their surface (specular reflection). If they scatter incident direct light they appear more matt (completely diffuse reflection involves an even scattering of light in all directions). A high proportion of specular reflection is characteristic of smooth polished surfaces. A high proportion of diffuse reflection is characteristic of more textured surfaces. It is thus the degree of smoothness or specularity of a surface that is the controlling feature in reflection. Despite the fact that on many occasions we talk about a mirror as if it is only specularly reflective or about a material such as grass as if it is entirely diffusely reflective, in fact all real surfaces display mixed characteristics. An enormous range of reflective optical properties therefore exists. Polished stone surfaces, for example, reflect light specularly and diffusely. Although they can be made very smooth, the proportion of incident light that they reflect specularly is not as high as with an ordinary mirror and the images they reflect are therefore less bright.

The issue is complicated further because the precision of a reflected image in a polished surface depends first on the specularity of the surface at a particular angle of view (many materials exhibit a range of specularity across different angles of view and are more shiny when viewed obliquely) and, second, on the strength of directionality of the incident light (glossy surfaces appear matt in completely diffuse light). The perceptibility of a reflection in a surface on the other hand depends on the proportion of reflected light that is diffuse (reflected images in black surfaces are clearer than those in white surfaces because the specularly reflected light is not accompanied by diffusely reflected light). Depending on the relative proportions of specularly and diffusely reflected light scattered by a surface, a reflected image or the surface itself is visible (Steane 1999).

REFLECTED IMAGES AND THE LOCATION OF SURFACES: THE SPECULAR THRESHOLD

The main hall of the Casa del Fascio is an interior where the use of polished planes of stone helps to animate a relatively straightforward spatial arrangement. Within a space that remains calm, airy and austere, planar stone surfaces gleam with reflected images of various degrees of precision and visual weight, ribbed glass surfaces glimmer or recede into darkness, flecked red granite walls sparkle in the half-light, and stainless steel handrails introduce sparkling highlights that contrast with the more visually stable surfaces of painted plaster. The viewer is thus denied a simple visual reading of the spatial enclosure. Space moves with the viewer: spatial boundaries transform with movement across the room as the viewer's location changes

10.17
**The creation of visual uncertainty: the view of the
main stair from the entrance area**

10.18
**The entrance area looking towards
reception**

with respect to the polished surfaces and the various sources of light
(Figures 10.17 and 10.18).

In the context of an analysis of Terragni's daylighting strategy it is
his decision to create a specularly reflective threshold to the building which
deserves further commentary, but first a brief set of remarks on the range
and distribution of surface finishes will indicate how the entry sequence is
composed in material terms. The location of a number of key interior stone
and glass surfaces will also be discussed. The effect of this disposition on
spatial experience will then be used to demonstrate how the space is
animated in light.

THE ENTRY SEQUENCE
The white, diffusely reflective marble finishes of the external ground surface
and front facade are succeeded by darker polished stone surfaces in the
single storey entrance area (black on the ceiling and end wall, rose-beige on
the columns and floor). The double-height hall has the same rose-beige
polished marble floor, though laid in smaller tiles, but its walls and gallery
ceilings have much more diffusely reflective painted plaster surfaces that are
light in colour. At the back of the building, further rose-beige polished marble
surfaces line the walls of the secondary stair and the rear wall of the hall.

THE USE OF STONE SURFACES
The most remarkable surface of an unusual interior is a continuous plane of
polished black marble that runs across the ceiling of the single storey
entrance area. (Once almost glass-like in appearance, it is now much less

10.19
Edge of the main stair — here the original glossiness of the black marble surface has been retained

shiny after suffering serious fire-damage.) To one side of the space it also lines the wall. To the other it extends across the entire underside and edge of the main staircase (Figure 10.19).

THE USE OF GLASS SURFACES

Balustrades to the main stair and the galleries around the hall are all of clear glass, as are panels that line the columns at the rear of the hall. (A curving glass balustrade introduces a particularly spectacular distorted reflection at the landing of the stair.) Reception offices at ground and first floor have screen walls composed of opaque and clear glass panels within brass frames that give this corner of the building a greater transparency. Elsewhere this is not the rule for interior openings: in general, translucent rather than transparent glazing is constructed of panels of glass-block or ribbed diffusing glass. At one side of the hall, for example, a columnar glass-block screen wall masks a group of ground floor offices. On the whole, doors to offices and other secondary spaces are not glazed and it is worth noting that Terragni decided that only the building's most important meeting room should have a strong visual connection to the hall. This was via a long full-height window located immediately above the glass-block screen. (In another change to the original arrangement opaque glass has now replaced the clear glass in this location.)

VISUAL CERTAINTY: THE IMPACT OF REFLECTED VIEWS ON SPATIAL AND TEMPORAL DIVERSITY

From the outset, visitors to this building have noted how unexpected views of the world outside are brought into the interior. This is because most of the polished wall and ceiling surfaces are associated with sources of direct light from external window openings. The polished black marble ceiling and end wall of the entrance are thus ideally located to bring extensive, if dark, views of the square and cathedral deep into the foyer (Figure 10.20). These reflections would have had a particular clarity in the glossy black surface. In the floor the reflected images are more blurred and less easy to perceive (the floor is less smooth, pale in colour and has a slightly undulating surface). While the reflections in the ceiling may be compared to those in a still, dark pond, those in the floor give the impression of a gently moving surface of water (Figures 10.18 and 10.21).

Several of the photographs in *Quadrante* describe the 'doubling' of space brought about by Terragni's decision to locate planar glossy surfaces immediately adjacent to windows. Early photographs of the entrance, for example, convey the visual release upwards into a dark doubled volume that is the ambiguous spatial consequence of a large-scale reflection. 'Doubling' of space also occurs at the main staircase because the marble wall surfaces, although light in colour, are viewed at very oblique angles on entry or exit (Figure 10.16).

10.20
Despite damage to the surface, reflections of the centre of Como are still perceptible in the end wall to the entrance area

10.21
Reflections in the floor are more perceptible when viewed from a more oblique angle — ground floor viewed from entrance area

Spatial boundaries are thrown into question perceptually where this kind of extension into virtual space occurs. Here the ever-changing perceptibility of the reflections in the planes of polished stone ensures that spatial definition is undermined but not completely negated. Yet the degree to which any particular reflected image qualifies spatial enclosure, and thus influences visual appearance, depends on the colour and surface texture of the surface in which it is seen, the size and location of the surface with respect to the viewer, and whether the image itself is in movement. Only a more in-depth analysis of detailed design decisions can therefore illustrate the level of coordinated decision making that this design strategy requires.

In the main hall of the Casa del Fascio, Terragni decided that the clearest and most spatially significant reflections should be those in the untouchable surface of the ceiling. Highly polished ceilings of this kind are relatively unusual, possibly because it is visually uncomfortable to have the world upside down above one's head, but in this case the toned down character of the reflections qualified the definition, and therefore 'weight' of this other world. The reflected images in the black surfaces of the stair are

10.22
Reflections in the floor are less perceptible when viewed from a less oblique angle — ground floor viewed from first floor

less easy to interpret in spatial terms. They matter temporally, however, providing the viewer with rapidly scintillating flashes of movement rather than glimpses into a dark mirror. This is also true of one of the more unexpected views to be gained in the hall. This turns the world round the other way, allowing movement in the neighbouring street to subvert the predominantly calm mood of the interior. From the first floor balcony above the entrance, a reflection of the traffic streaming by the building is captured in the glass balustrade of the gallery opposite.

VISUAL STABILITY: DISCOVERING SPACE AND LIGHT THROUGH MOVEMENT
In less glossy surfaces, such as the floor, stronger reflected images are only apparent at an oblique angle of view in light that is sufficiently direct. (Less highly polished stone surfaces tend to have a lower specularity when viewed from angles close to the normal, diffuse light lowers the perceptibility of reflected images.) This means that the entrance area, when viewed from a distance, does appear to be doubled in the floor, but that the main hall floor surface, when viewed from directly above under the more diffuse light of the central skylight, is more visually stable. Yet when the floor is viewed from an oblique enough angle, reflected images can still be seen (Figures 10.21, 10.22 and 10.23).

Similarly, the less highly polished vertical surfaces of the columns in the entrance area offer a different appearance depending on the angle at which they are viewed. Seen from the sides of the hall they appear matt, seen more obliquely from the back of the hall their shininess is apparent. When viewed straight on they are thus able to provide a visually stable frame to other more distant reflections in the floor, walls and ceiling, but when viewed more obliquely they help to subvert the primary spatial definition (Figure 10.24).

10.23
'Glaces à Répétitions': the reflected image of the floor in the ceiling is mirrored again in the floor — view towards the main entrance from the rear of the hall

Honed stone surfaces (surfaces that are smooth but not polished) change in appearance in a more dramatic way at different angles of view. When viewed straight on, the honed black marble used on the exterior walls of the bathrooms appears completely matt. When viewed more obliquely, reflections suddenly become visible because such surfaces exhibit an abrupt change in specularity with angle of view (Figure 10.25). Movement along the wall between the front and the back of the building is thus able to be captured in this surface. For the viewer who looks at it from either end of the building, a view is offered that anticipates the space, light and activity in the space beyond.

THE BLACK CEILING AND THE LIGHT-COLOURED FLOOR: LIGHT DISTRIBUTION
Black is an unusual choice for a soffit because ceiling surfaces usually play an important role in distributing light within interiors. Here, the fact that little of the light that hits the ceiling is reflected diffusely helps to ensure that the directionality of the light in the entrance area is upwards rather than downwards, and endows the lighting in this area with a theatrical quality comparable to that of theatre footlights. It also ensures that, despite being

10.24
The shiny surface of a column in the entrance area is not apparent when viewed straight on

10.25
Reflections of the hall may be seen in the honed black marble surface of the wall to the bathroom when it is viewed at a glancing angle

well lit from the outside, the front area of the hall is not significantly lighter than the area of the hall below the light-well. Relatively even light levels of this kind mean that areas farther from the front facade do not look gloomy and that, overall, the lighting has a soothing, slightly subdued quality, which helps to make the interior a visual refuge from the much brighter light outside (Figure 10.25).

The glossiness of the floor and ceiling surfaces ensures, however, that some of the light that enters through the front facade can be redirected deep into the interior, and that in the process they become relatively bright secondary light sources for those viewing the entrance area from the back of the hall.

MATERIAL SPECIFICATION, SPECULARITY AND LIGHTING DIVERSITY

> In this building all is derived from the outside, from the square, from the air; and everything suddenly moves and becomes external once more; mediating a rich series of episodes beginning with the substitution of the old means of entrance ... and by means of the very frequent and surprising use of that glass everywhere, for example in sheets on the walls and on the ceiling, by means of which even in the most internal parts of the building elements of the most distant views remain visible, among them those of the summit of Brunate.
>
> (Bontempelli 1936: 366–8)

In this building Terragni grasps what the architectural exploitation of specular reflection entails in design terms. Decisions regarding the overall cross-section of the building are coordinated with the size and position of the building's windows, the specification, location and juxtaposition of material surfaces, and with typical patterns of movement, to create a building whose reflections contribute as much to the spatial ambience as to the play of light. It is a solution in which a combination of diffuse and direct incident light endows the visual environment of the hall with a peculiarly calm and aqueous quality, enhanced by the gleaming highlights that result from a carefully considered distribution of polished and matt surfaces. The relatively even light levels within the interior are a further critical aspect of this lighting strategy. High enough to ensure that the building is adequately lit, and that looking out of the entrance doors is not an uncomfortable experience, they are nevertheless sufficiently low to ensure that a whole range of reflections are perceptible to the occupants. Terragni's intention was to avoid the harsh brightness contrasts produced by strong sunlight and shadow. To a large extent only diffuse and specularly reflected light are therefore brought into the heart of the building. In a climate where restricting the penetration of sunlight is critical to achieving thermal comfort, reflections rather than shadows and light ensure the luminous diversity of the interior.

Conclusions

Bontempelli's description of the Casa del Fascio emphasises again, if somewhat ambiguously, the significance of movement to the richly poetic interplay choreographed by Terragni between the interior and exterior of his building (Bontempelli 1936: 366–8). Although it is not a very precise summary of the building's lighting strategy, the comments usefully focus attention on the dynamics of the environment as it is experienced. Throughout this chapter the importance for lighting diversity strategies of three different kinds of movement has been underlined: first, the movement of the viewer across space, second, the movement of the sun with respect to building apertures, and third, the movement of reflections and reflected light in spaces where highly specular materials are employed. All the natural lighting strategies discussed here have in their different ways reciprocated such patterns of movement while at the same time taking account of the visual requirements of particular spatial sequences and settings and the quality and strength of natural light available.

By paying close attention to the visual environments he was defining, Terragni articulated the role he wished the Casa del Fascio to play in the urban topography and public life of Como with great eloquence. Its main hall is a space that provides a strong visual link to the city and is yet of a quiet, contemplative character whose 'austere vitality' simultaneously states its connections with, yet removal from, the more everyday world outside. As in the projects discussed previously, it is a strategy which demonstrates the extent to which a concern for visual diversity can inform and enrich architectural thinking by underlining key spatial relationships, guiding transitions, establishing important thresholds and inviting particular patterns of movement.

References

Bontempelli, M. (1936) *L'Avventura Novecentista*, reprinted 1974, Florence: Vallechi.

Neufert E. (1936) *Bauentwurfslehre*, Braunschweig, Wiesbaden: Friedrich Vieweg und Sohn.

Steane, M. A. (1999) 'Lustre. Specular reflections in buildings: An investigation of optical and perceptual issues', unpublished MPhil dissertation, University of Cambridge.

Terragni, G. (1936a) 'La costruzione della Casa del Fascio di Como', *Quadrante*, October 35(36): 5–27.

Terragni, G. (1936b) 'Relazione tecniche', *Quadrante*, October 35(36): 38–55.

Walker, B. (1998) *Gabriel Poole — Space in which the soul can play*, Australia: Visionary Press.

Chapter 11

Daylight perception

Katerina Parpairi

Introduction

Daylight's primary function is to activate the visual world around us, so that we experience spaces, colours, emotions, etc. This is especially obvious in the arts, whether in painting, sculpture or architecture; light reflected on surfaces links the visual environment with human feelings, creating mood. Famous architects and artists have used light in order to convey messages to, and most importantly stimulate emotions in, the observer. Architectural masterpieces such as the Hagia Sophia of Constantinople or the chapel at Ronchamps (Figure 11.1) are fine examples of how daylight is controlled by the building envelope and reveals the form of the building with its dynamic character.

However, the knowledge of the dynamics of daylight and its effect on the user of a space was slowly lost. Aesthetics and function were

11.1
Le Corbusier's 'light wall' at Notre-Dame-du-Haut, Ronchamps, France (1954)

separated since the age of Enlightenment due to the development of the mathematical approach. Daylighting, even though once inseparable from the practice of building design, began to be regarded as anachronistic since the 1930s due to the availability and practicality of electric lighting (Figure 11.2). This was mainly based on the provision of predictable and constant illumination levels as opposed to the natural and unpredictable variations of daylight. However, the recent escalation of costs and the parallel rising concern about depleting resources, gave rise to the development of lighting design techniques in order to reintroduce natural lighting in the buildings.

11.2
Deep plan offices of the 1970s provided the users with a completely artificial environment, excluding daylight and natural ventilation

In modern times when it comes to designing with daylight, designers, engineers and researchers refer to numerous codes and standards (e.g. CIBSE 1994), which elaborate primarily the quantitative recommendations but are lacking in offering overall qualitative advice. We know how to achieve a certain recommended illuminance level on, for example, the workplane of a side-lit office but we know little about how this office will appear to its user.

The preoccupation of scientists with the physical aspects of light have led to designs which have to correspond to specific numerical light levels, irrespective of the interior or exterior view of the user, their position in the space, not to mention, of course, the opportunity and freedom they could be given to alter the lighting of the immediate workspace. Quantitative analysis is reasonably straightforward. A designer simply chooses a convenient prediction method (e.g. daylight factor calculation) and applies it to the building. The resulting values are then compared with recommendations in codes or specifications and the quantitative acceptability of the design is settled unambiguously. This raises the question: is it possible to understand the aesthetic and emotional aspect of daylight — referred to as quality — through numbers?

Daylighting quality is a very ambiguous term. Qualitative design is much more personal in terms of architectural design intentions and user preferences. Boyce summarises the current state of knowledge with the following:

> we really have no idea of what produces good lighting ... At the moment, it most frequently occurs at the conjunction of a talented architect and a creative lighting engineer, neither of whom is given to slavishly following numerical criteria.
>
> (Boyce 1998: 74)

Indices and numbers cannot fully account for issues of aesthetics and emotions stimulated in the users of a space. Daylighting does more than just make things visible. It is still, first and foremost, a perceptual issue (Villecco *et al.* 1979).

Visual perception

When we look at the world around us, the parts we look at light to our eyes, which comes to them from a source of illumination (Figure 11.3). An elaborate series of mental processes take place, which convert the visual pattern in the brain into the perception of the world as we know it (Lee 1997). This visual pattern is not static; it continuously flickers and moves, due to the saccadic movements taking place when the eye is searching (Gregory 1990). On the other hand, even though the eyes flicker, we perceive the world in comparative constancy and stability while the perceptual system is geared to understand and respond appropriately to frequent change.

A clear example of the sophistication of the perceptual process is brightness perception. The eye has a built-in adjustable sensitivity: a given amount of flux stimulating the retina under one set of circumstances will produce a different sensation of brightness from the same amount of flux under another set of circumstances. So a certain brightness seen in dark surroundings will appear much brighter than if it was seen in a brighter environment because of this adaptation-sensitivity mechanism (Hopkinson *et al.* 1966). The phenomenon of brightness is fundamental to the entire experience of viewing the world. It is a basic component of visual perception and as part of light theory is intricately connected to the design of the architectural environment. In many cases, engineers, architects and lighting technicians have little understanding of the relationship between the amount of light and the perception of brightness.

From this very simplified and short introduction on perception, two points become apparent. First, the eye is habituated to continuous change, thus standardisation in terms of daylight design recommendations is not necessarily the answer to a successful environment. Second, due to

11.3
Interior view of the Darwin College Study Centre, Cambridge, UK, designed by Jeremy Dixon and Edward Jones (1994), where a source of illumination is reflected light from the river outside bouncing on to the ceiling

the eye's adaptation mechanism, each scene is viewed differently depending on the relationship between the brightness of the focus point and its surroundings.

> In and of itself, the measured surface luminance of an object does not determine per se how bright the object will appear as perceived by the conscious mind. That perception, like any other, is influenced by a host of related factors, all of which combine to determine the perceived brightness of the object. This is why the specification of single-valued numerical criteria for the luminous environment, such as minimum footcandle levels, gives no guarantee whatsoever that the resulting environment will be perceived as bright or cheerful, pleasant or appropriate. All such judgements are based on holistic, complete perceptions involving the entire visual field, as well as expectations and prior experience.
>
> (Lam 1977: 40)

The complexity of the perceptual mechanism leads us to the conclusion that it is rather unlikely to be able to quantify daylight perception and thus design the perfect daylit interior. Since daylight perception is a highly personal process, the best way to explore this process is by studying the observer and their response to different environments. The importance of human assessment is based on the fact that man adapts to his surroundings, in contrast to a measuring instrument, and this is often a determining factor in the success of a design. The performance of the visual system is not simple, so if one takes advantage of man as a measurer of his environment, one might find an underlying simplicity despite the sophistication of the sensory system. The environment is not perceived as complex: it is our everyday environment and the user will always judge it in simple terms.

Spatial and daylighting diversity: user preferences

Asking the users of a space to describe their perception of daylight through a questionnaire (Figure 11.4) and then statistically comparing their responses with physical measurements is a simplified description of the procedure adopted in behavioural science and experimental psychology, through psychophysical (Flynn *et al.* 1973) and Post Occupancy Evaluation research (Heerwagen and Loveland 1991). The main framework of these types of research approaches into daylighting quality is that users of a space are employed as the measurers of their environment.

An overview of work in daylighting studies has revealed two major gaps. On the one hand, the process of relating physical aspects of the luminous environment to its psychological effects has remained largely unfulfilled. On the other hand, very few studies have been performed in the

11.4
View of the Martin Centre library, Cambridge, UK, where students were asked to fill in a questionnaire on daylighting perception

field, due to the unpredictable, ever-changing daylight conditions. Most psychophysical studies have taken place in laboratory studies with artificial lighting. The real world however, is a complex place physically, socially and culturally, and laboratory studies miss out this complexity.

In response to the identification of these gaps, a recent study (Parpairi 1999) used the existing methodology in order to address the issue of how the users of a real space perceive daylight. This research started from the user end of the issue and worked backwards to identify what design aspects affected him/her in judging a space to be well or badly lit. Some of the major findings which are discussed here illustrate clearly the basic point which is emerging: the way occupants perceive daylight quality is significantly affected by non-lighting and non-quantifiable variables, which will subsequently reveal the importance of daylight diversity in terms of user comfort and satisfaction.

The case studies selected were three Cambridge libraries, compiling as much diversity as possible:

1   Middleton library (Department of Architecture and History of Art library), housed in an 1839 terrace house — the architect is unknown (Figure 11.5(a));
2   Darwin College Study Centre, designed by Jeremy Dixon and Edward Jones and completed in 1994 (Figure 11.5(b));
3   Jesus College library, designed by Evans & Shalev and completed in 1995 (Figure 11.5(c)).

As shown in the images, seven reading positions (Figure 11.5) in total were tested under a sunny sky in summer and winter, and under an overcast sky in autumn. The Middleton library experiments took place in two

a

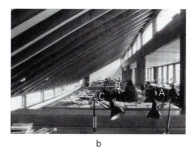

b

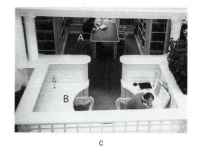

c

11.5
(a) Interior of Middleton library with positions A and B noted on the image; (b) interior of the Darwin College Study Centre's upper floor with positions A, B and C noted on the image — position A has abundant views towards the river Cam, but has a significant glare problem; (c) interior of the Jesus College library bay area (position A) and carrel (position B) — position A is surrounded by bookstacks and has a beautiful view out towards the College gardens; the carrel area has very high reflectances in the field of view

identical rooms, one daylit, and the other with the artificial lights on. The stability of the sky conditions was also confirmed by the measured data. It was decided that the subjects who participated in the experiments were to be non-expert in the field, since trained subjects are apt to become too sophisticated and critical (Hygge *et al.* 1996). They were a consistent group of 26 randomly selected students, who used the libraries regularly and who were thus familiar with the spaces and their daylighting environment. In order to obtain information about the daylighting experience of the subjects in the libraries, a survey methodology was developed which simultaneously recorded the various subjective feelings through a questionnaire and the local daylighting conditions related to an individual subject through environmental monitoring (physical measurements and photography).

By statistically assessing the subject's questionnaire responses and the physical measurements through statistical analysis, this research identified two major groups of parameters which affected the subjects' perception of daylight quality The first is named 'quantifiable' parameters since they can be measured with instruments, and the second is named 'architectural and personal' parameters since they are not quantifiable.

It is obvious, of course, that physical parameters such as illuminance and reflectance are important in terms of how the user perceives daylight, without of course being the sole parameters affecting his perception.

The significant correlations between all the questionnaire rating scales (bipolar n-category rating scales such as 'pleasant–unpleasant') and the log of the illuminance levels for each subject were found to be in some cases significantly strong (as high as 0.75 — regression analysis indicated that the logarithmic relationship explained up to 48.5% of the variance on the rating scale). Thus, illuminance is an important factor in a space, and the fact that it is included in several codes and standards as a design criterion is justified (Figure 11.6). However, from the statistical analysis results it becomes apparent that simply the illuminance level, and thus target or recommended levels, cannot alone ensure a satisfactory daylighting environment. Furthermore, the logarithmic relationship indicates that above the

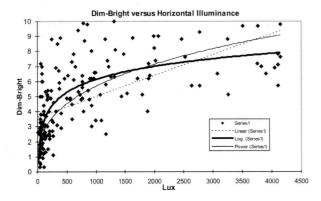

11.6
Scatter diagram of 'dim–bright' voting scale against horizontal illuminance. In this case the logarithmic relationship explains 45.4% of the variance on the dim–bright scale

recommended threshold (500 lux), the sensation levels out. Thus, this work indicates that by raising the levels one does not make a space seem brighter or more satisfactory.

In terms of surface reflectances, results show that the higher they are, the brighter the space appears. There is of course a limit to this, since when reflectances are too bright (e.g. Jesus College library's position B, Figure 11.5(b)) or when surfaces have specular finishes (e.g. Middleton library, Figure 11.12), then discomfort glare becomes an issue which is often mentioned by respondents.

Illuminance ratios (the recommended 10:3:1 is rarely achieved in real interiors) on the other hand as well as the Daylight Glare Index, which are both discussed often in bibliography, were unsuccessful metrics to predict perception in the daylit environment. In particular, the Daylight Glare Index was found to over- or under-predict the glare sensation — a conclusion which is in line with other studies (Hopkinson 1971) — for several reasons, of which the most significant is that it is not possible to take into account the view content and appearance or design of windows.

The statistics prove that illuminance levels and reflectances are important factors in design and thus codes and standards are not to be ignored. However, users in everyday life do not simply judge a specific task such as 'satisfactory daylighting level for reading', but rate the whole spatial experience of daylight in an interior. For this reason the subjects in this study completed a questionnaire which included rating scales and some open-ended questions which helped the author to identify other parameters which could not be measured electronically. These were named 'architectural and personal' parameters since they refer either to the design of the space or to psychological and emotional parameters.

Views out

It seems bizarre that one has to note the importance of having a view out from a building. However, building design practice, due to the extensive use of artificial lighting, often ignores its importance, and thus fewer and fewer users have the privilege of connecting to the outside world (this is discussed in a broader context in Chapter 4 of this book by Nick Baker). Although totally windowless buildings are rare, many buildings have interior spaces and below-ground levels that do not have windows. As many as 50% of the users working in modern high rise office buildings do not have access to windows, and current interior design may have increased this number by producing de facto windowless spaces (Figure 11.7). There appears to be a widespread consensus in the research community that users do not like to work in windowless offices (Collins *et al.* 1990; Heerwagen and Loveland 1991).

Heerwagen (1990 and 1998) summarises clearly the four general benefits of windows and natural light.

11.7
**'Window blinds down and lights on':
a usual working environment for VDU
users**

1   Access to environmental information — windows give us information on the time of day, seasonal changes in vegetation, weather and other forms of environmental data, which helps to maintain our biological cycles.

2   Access to sensory diversity — change, whether slow or rapid, great or minute, is a basic characteristic of the natural world. Sensory diversity is fundamental to perception and the constantly changing nature of daylight satisfies the biological need of the mind and body for this change. Our indoor environments are largely devoid of sensory change, and deliberately so. For many users, windows may provide the only access to changing levels of sensation.

3   A feeling of connection to the outside world — windows provide us with access to events and situations in the world beyond our walled boundaries.

4   Restoration and recovery — one of the benefits is 'psychological relief'. This also depends on the view content. Those predominately of a natural content are more restorative than views consisting of built spaces without nature.

Returning to the daylight study of libraries (Parpairi *et al.* 2000), the questionnaire responses illustrated clearly that subjects enjoy having a view out of their working area. They not only welcome the fact that they have the opportunity to look out, but they are also willing to accept high levels of glare in order to enjoy this privilege, as in the case of Darwin's position A, where sunlight during winter days penetrates the interior through the south-east facing windows (Figure 11.8). Architects should therefore be encouraged to incorporate windows for view as well as daylight, within proximity of workstations. Having a workplane-based task does not negate the need for visual diversity. The least successful locations were reported from occupants

11.8
View of position A in Darwin College Study Centre — this library has a wonderful view out over the river Cam

11.9
View of position B in Jesus College library — notice how most of the field of view of the subject is white

with no view out (e.g. Jesus' position B where subjects were less tolerant of the 'whiteness' surrounding them probably because their eye had no resting point) (Figure 11.9).

There is no need to consider a view out as compromising visual comfort. One can achieve the latter (e.g. with the use of advanced daylighting systems) while at the same time offering the user of the space the opportunity for relaxation and stimulation. In other words, users prefer the variability and diversity of a view out of a window, even though this could create glare problems, and thus undermine visual comfort in a technical sense.

Sunlight and adaptation

The issue of adaptation was brought forth in this study with the response of subjects to different glare situations. Darwin's position A under sunny conditions was rated on the 'glary–non-glary' scale as uncomfortable due to the sun penetration and the lack of shading devices along the south-east facade of the building (Figure 11.10). The second most glary situation was perceived under both sunny and overcast sky conditions in Middleton due to the annoying veiling reflections on the table from the overhead lamps (Figure 11.11). The first was due to natural phenomena. The second was due to artificial means of illumination. What were the comments made by the subjects?

In Darwin's case the subjects' comments were not as harsh as might have been expected (Figure 11.10). The extended tolerance may be due to the fact that the subjects acknowledge having the option to change seating position. In other words they feel less constricted in this situation:

> The worst activity from the sunlight point of view is when occupants are restricted in their movements to rigid seating or standing positions with limited freedom of viewing directions. The

11.10
View of position A in Darwin. Glare is dramatic due to the lower sun angle. This subject noted that he still prefers this position to others in the space

11.11
The subject sitting in position B in Darwin, changes his posture in order to avoid the patch of sunlight from the north-east clerestories in summer

more freedom the occupants have to change the direction of their view or position in relation to the sun, the less negative the effect of the sun may be.

(Ne'eman 1974: 163)

This adaptive response was also observed under other occasions, when everyday users of the library spaces changed locations and positions depending on the sunlight penetration. Furthermore, there were cases where users would simply change their posture in order to either avoid a sun patch or to welcome more light towards the end of the day (Figure 11.11). Another reason for their tolerance of this daylit environment would appear to be the fact that the view out of the windows on the south-east side is so pleasant, as mentioned earlier.

However, in Middleton library the subjects are much less satisfied with the lighting environment, with responses such as: 'The overhead light reflects off the shiny surface of the table and hurts my eyes' (Figure 11.12). The lower levels of tolerance in this room are explained by the fact that they accept discomfort from natural sources (daylight) more easily than from artificial ones. Another possible reason could be that the subjects in this situation have no alternative seating position (except for two armchairs) in order to avoid the annoying veiling reflections. Irrespective of their seating position they are not able to avoid the desktop specular reflections.

It would be expected that people working in spaces where they are subjected to glare for a longer time, or where they cannot turn away from the glaring light source, will have less tolerance to glare than people working in spaces where they can, by moving their working position or attitude, obtain alleviation from the glare.

(Hopkinson 1970: 101)

11.12
View of position B in Middleton library. Glare for the occupants is primarily due to artificial lighting

From this anecdotal evidence it becomes obvious that adaptation is a normal activity for the users of a space, and the more opportunity they are given the more satisfied and tolerant they are of visual comfort. Baker and Standeven (1996) note that when there is no adaptive opportunity, any departure from neutrality will cause stress or dissatisfaction. This is the case in the Middleton library. For this reason it is important to allow for individual control (e.g. individual adjustable desktop lighting, controllable shading devices, etc.) in order to encourage the interaction of the user with the daylighting environment. These conclusions have also been arrived at in other studies addressing either thermal comfort (Baker 1996; Baker and Standeven 1996) and in the interior environment in general (Heerwagen *et al.* 1992; Bordass and Leaman 1997).

Diversity

The eye seeks information primarily manifested by contrasts, highlighting obvious differences of brightness and/or colour (Liljefors 1997). Several studies have discussed the importance of luminance to lighting perception, but there has been no satisfactory representation of luminance except in terms of recommended ratios. However, the resulting ratios in real-life daylit interiors are far greater than the recommended limits of 3:1 and 10:1 and yet subjects were frequently satisfied with the conditions. Furthermore, ratios are typically based on averaging luminances, whilst what is of interest is to assess luminance variability and thus luminance contrast.

The research work presented here developed and employed a new method of representing the daylight distribution at a point in the room (Parpairi *et al.* 2002), where 360° luminance measurements were taken in a horizontal plane at eye level at the position of the subject (Figure 11.13). Based on research of the eye scanning amplitude, the log of the differences of successive luminances was taken over different acceptance angles (11.25°, 22.5°, 45° and 180°). These differences were then summed up, and

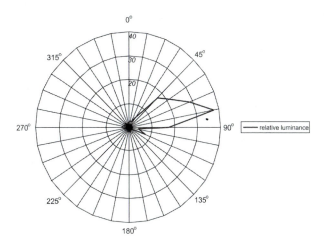

11.13

Polar diagram of luminance measurements in Jesus library's position A, at 11.00 a.m. under a summer sunny sky

the maximum sum is chosen to represent the highest degree of variance for that specific angle ('LD index'). By performing statistical (regression) analysis with the subjects' responses from the questionnaires, it was concluded that diversity, meaning in this context the luminance variability, is highly appreciated by the subjects in the library spaces. The more diverse the luminances in the field of view, the more 'pleasant', 'cheerful', 'bright', 'radiant', 'clear', 'visually warm' and 'strong' the space was reported to appear. Interestingly, even the task lighting appears better to the subjects when there is this visual diversity.

In Jesus' position B under the sunny sky, subjects are not as satisfied as would be expected from the daylight levels (with mean illuminance levels exceeding 1500 lux) and 42% of occupants would switch on the desktop lamp (Figure 11.14). This seems counter-intuitive when we look at the equivalent sunny sky responses (Figure 11.11) for Darwin's position B (with mean illuminance levels below 1200 lux), where none would switch lights on even though the illuminance levels were lower than those recorded in Jesus College library. The reaction is probably explained by the fact that most general comments made pointed out that subjects found Jesus College library daylighting too white and glary compared to the softer artificial lighting. This seems to suggest that artificial lighting is not only used to counteract problems of insufficient daylighting but also to ameliorate the lighting quality in the field of view. It is possible also that due to the lack of view (the monotonous and constricted field of view was noted by a number of subjects), and thus limited stimulation, subjects may unconsciously be using artificial lighting to create interest and help focus their attention.

From the above, it can be argued that in order to satisfy the subjects in terms of brightness, there is no need to have very high illuminance

11.14
View of position B in Jesus College library. The subject has switched on the light in order to 'soften' the light around him and to create a focus for his work

or luminance levels. Diversity and not intensity is required to satisfy the subjects to a certain extent. This parameter relates closely to designing interiors with variable reflectances, not only in terms of different wall reflectances, but also by the changing conditions from a high reflectance wall, to a bookstack, to a window, thus creating visual interest as well as luminance variation.

Control

A general conclusion from the subjects' opinions is that control is welcome, particularly at an individual level. Desktop lighting is often mentioned as a positive aspect of the space and this fact becomes even more obvious when individual desktop lighting is absent, as in the case of the Middleton library (Figure 11.5(a)). It is difficult for users to request changes in general lighting arrangements.

The users of a space appear, from our research, to enjoy interacting with their environment. Subjects were willing to use blinds, or desktop lights as needed. In Darwin's position A under a sunny summer sky most subjects mentioned that the position is very pleasant due to the view, but there is need for blinds to control glare (Figure 11.10). Furthermore, variety in an interior in terms of optional positions is also described as a type of control, and so in Darwin College Study Centre the subjects mention this opportunity and also the fact that they change seating positions if they feel the need to.

We are aware that satisfaction is strongly connected with individual control and adaptive opportunity. Thus, it is important to allow for user-friendly controls in a workplace (blinds, curtains, desktop lighting), whilst at the same time providing a variety of seating locations and positions if possible. The user will subsequently have the option to sit closer to the window and counteract possible visual comfort problems by drawing the blinds, or sit farther away from the window wall and still enjoy a view out, or rotate position to improve view or reduce glare. Diversity, flexibility and interaction are thus the key words in a daylit library interior.

Conclusions

This chapter revealed, through research in a series of case study libraries, the correlations between some physical daylight measurements and the subjects' descriptions of their perception of their luminous environment. Initially, it confirms that certain basic physical factors significantly affect user response, while others which are traditionally considered significant by international codes and standards were less significant.

The research moved on to illustrate an important point: there are a series of non-quantifiable parameters, described as architectural and personal, which are highly important aspects of daylight design and which affect the user's daylight perception significantly. These are simply described as: view out, adaptation, diversity and control.

The users of an interior daylit space enjoy having options. They enjoy the opportunity to change seating position, to switch lights on and off, to draw curtains/blinds, to look out of a window and sense the hourly and seasonal changes of climate. They also prefer to have a diverse environment and dislike uniformity in terms of reflectances. If they feel that they have the freedom to control their environment they are also more tolerant of the extremes of natural daylight.

What does this mean for design? The fact that quantitative parameters are only part of the 'equation' and that daylighting quality cannot be defined in simplistic terms, should not be seen as a drawback. In fact, the opposite is argued here. Opportunities open up for the design team as a result of an awareness of the significance of luminous diversity, responding to daylight perception as a motive, as opposed to being led solely by objective (yet sometimes erroneous) criteria, thus increasing design freedom. How designers exploit the conclusions of behavioural research will give a unique identity to the buildings they design, based on their awareness of, intuition for and sensitivity to the future users.

References

Baker, N. V. (1996) 'The irritable occupant: Recent developments in thermal comfort theory', *Architectural Research Quarterly*, 2(2): 84–90.

Baker, N. V. and Standeven, M. (1996) 'Thermal comfort in free-running buildings', *Energy and Buildings*, 23: 175–82.

Bordass, B. and Leaman, A. (1997) 'From feedback to strategy' in *Buildings in Use '97: Results from the PROBE Research Project*, London: CIBSE.

Boyce, P. R. (1998) 'Lighting quality: The unanswered questions', *CIE Proceedings of the First CIE Symposium on Lighting Quality*, 72–84, Vienna: Commission Internationale de l'Eclairage.

CIBSE (1994) *Code for Interior Lighting*, London: The Chartered Institution of Building Services Engineers.

Collins, B. L., Fisher, W., Gillette, G. and Marans, R. W. (1990) 'Second-level post-occupancy evaluation analysis', *Journal of the Illuminating Engineering Society*, Summer: 21–44.

Flynn, J. E., Spencer, T. J., Martyniuk, O. and Hendrick, C. (1973) 'Interim study of procedures for investigating the effect of light on impression and behavior', *Journal of the Illuminating Engineering Society*, 3(1): 87–94.

Gregory, R. L. (1990) *Eye and Brain: The psychology of seeing*, London: George Weidenfeld and Nicolson Ltd.

Heerwagen, J. H. (1990) 'The psychological aspects of windows and window design', 269–80, *Proceedings of the Twenty-First Annual Conference of the Environmental Design Research Association (EDRA 21)*, Champaign-Urbana, IL: Environmental Design Research Association.

Heerwagen, J. H. (1998) 'Of light, time and space: Lighting quality and green building design', 211–18, *Proceedings of the First CIE Symposium on Lighting Quality*, Vienna: Commission Internationale de l'Eclairage.

Heerwagen, J. H. and Loveland, J. (1991) *Energy Edge, Post-Occupancy Evaluation Project, Workspace Satisfaction Survey*, Final Report (WA 98195), Seattle: University of Washington.

Heerwagen, J. H., Loveland, J. and Diamond, R. (1992) 'Coping with discomforts', 3(1): 3041–7, *Proceedings of the Biennial Congress of the International Solar Energy Society, 1991 Solar World Congress*, New York: Pergamon Press.

Hopkinson, R. G. (1970) 'Glare from windows', *Construction Research and Development Journal*, 2(3): 98–105.

Hopkinson, R. G. (1971) 'Glare from windows — 2; What people say', *Construction Research and Development Journal*, 2(4): 169–75.

Hopkinson, R. G., Petherbridge, P. and Longmore, J. (1966) *Daylighting*, London: Heinemann.

Hygge, S., Löfberg, H.-A. and Poulton, K. (1996) 'A manual for post-occupancy evaluation (POE) and test-room studies', unpublished research report for the EU Joule II 'Daylight Europe' project, Denmark: Esbensen Consulting (coordinator).

Lam, W. M. C. (1977) *Perception and Lighting as Formgivers for Architecture*, New York: McGraw-Hill.

Lee, J. (1997) 'Before your very eyes', *New Scientist*, 15 March, 99(2073): 1–4.

Liljefors, A. (1997) 'Lighting and colour terminology', paper presented at a CIE discussion, Stockholm: Commission Internationale de l'Eclairage.

Ne'eman, E. (1974) 'Visual aspects of sunlight in buildings', *Lighting Research and Technology*, 6(3): 159–64.

Parpairi, K. (1999) 'Daylighting in architecture: Quality and user preferences', unpublished PhD thesis, University of Cambridge.

Parpairi, K., Baker, N. and Steemers, K. (2000) 'Daylighting quality through user preferences: Investigating libraries' *Proceedings of PLEA 2000, Cambridge*, London: James & James.

Parpairi, K., Baker, N., Steemers, K. and Compagnon, R. (2002) 'The Luminance Differences index: A new indicator of user preferences in daylit spaces', *Lighting Research & Technology*, 34(1): 53–68.

Villecco, M., Selkowitz, S. and Griffith, J. W. (1979) 'Strategies of daylight design', *AIA Journal*, September, 68–85.

Chapter 12

Exploring thermal comfort and spatial diversity

Abubakr Merghani

Introduction

People have always attained their desired thermal condition for living in a variety of ways (e.g. building a fire, opening a window, adjusting a thermostat, moving to a cooler/warmer location, etc.). The diversity of their responses to thermal stimuli could, therefore, be used to allow 'free trade' in environments suggesting that the regulation of natural variation in space and time could be more important than its total elimination — often at great cost (Markus and Morris 1980).

In hot-dry countries, movement between different living spaces is one of the most commonly practised behavioural adaptations, facilitated by the wide range and variation of spaces usually found in traditional buildings. Many people in such climates migrate within their buildings in both daily and seasonal patterns to take advantage of the various microclimates the buildings create.

Fieldwork was, therefore, conducted in Khartoum, Sudan, to investigate how people make use of the spatial diversity offered by the traditional single-storey courtyard houses which 75% of the urban population inhabit. The fieldwork monitored thermal conditions in four houses for six months (January to June 2000 — covering winter, spring and most of the summer season). Moreover, an observational study — conducted in 2000 and 2001 — noted space-use patterns and the main behavioural adaptation strategies. The objective of this investigation was to examine the direct/indirect implications of these adaptive strategies on the overall thermal satisfaction level as well as the spatial composition and energy consumption of residential buildings.

Spatial diversity as an adaptive opportunity

A building increases the available range of thermal zones so that people can select the microclimate most suited to their thermal needs. Buildings can, thus, be seen as a way to modify a landscape to create favourable new microclimates. Humphreys (1997) argued,

> when people are free to choose their location, it helps if there is plenty of usable thermal variety. Then they can choose places they like, which are suitable for the activity in which they wish to engage.

From ancient history to recent times people have always used available ranges in their shelters to adapt to local climatic conditions. Knowles (1999) describes how the location and form of the Longhouse Pueblo settlement provided its ancient residents with year-round comfort. The ancient settlement (c.1100 AD) is sited in a large, south-facing cave. The cave dwellers of Mesa Verde tended to migrate in and out of the cave in response to the north-south seasonal migration of the sun. This seasonal adaptation at Longhouse is complemented by the Pueblo's response to a daily rhythm as the sun moves from the eastern to the western sky, casting morning rays inside the west end of the cave and twilight rays inside the east end.

In the 1980s, Roaf monitored courtyard houses on the Persian Plateau. The rooms inside these houses offered varying thermal conditions and the occupants moved from one room to another at different times of the day. Roaf (1988) concluded that their movement was to a large extent motivated by the thermal profile of each space. Similarly, in a recent study, 10% and 11% of the subjects in cold and warm conditions respectively, moved to a more comfortable place (Heerwagen et al. 1991). Recently, Al-Azzawi (1996a and 1996b) studied such movements in courtyard houses in Baghdad and concluded that occupants' quest for comfort led to these ritual seasonal and daily migrations.

Similarly, Heidari (2000) conducted a one-week study of how six subjects used different parts of their house in Ilam (Iran). The subjects under study moved between a courtyard, Iwan (veranda), a family room and a basement. Throughout the study period they chose to occupy spaces which were within or close to their comfort zone. Accordingly, Heidari concluded that the subjects 'tended to actively seek out the most thermally comfortable spaces in the house during the course of the day' (Heidari 2000).

Clearly, similar examples of how people make use of the available diversity offered by their houses could be found all over the world. In Khartoum, the buildings and the use to which they are put have evolved together in such a way that it is no longer possible to say that the buildings are a direct response to the pattern of living nor that the way of life is determined by the buildings and the environmental conditions. Each is nicely matched to the other as remarked by Rodger (1974).

Psychological aspects of spatial diversity

The world is colourful and people enjoy this richness, nobody would commend a monochromatic world. And yet, a steady-state thermal environment is the prevailing standard for office buildings, schools and homes in many countries (Heschong 1979).

A constant temperature is maintained in order to save people from the effort and the distraction of adjusting to different conditions. And yet, in spite of the extra physiological effort required to adjust to thermal stimuli, people definitely seem to enjoy a range of temperatures. They frequently seek out extreme thermal environments for recreation or vacations (Baker 2000).

> The technology of heating and cooling aims ... to achieve a thermal 'steady-state' across time and a thermal equilibrium across space. In other words a constant temperature everywhere, at all times ... neither of these criteria is easy to achieve.
>
> (Fitch 1972)

Spatial diversity effects on thermal satisfaction and energy consumption

Many researchers have pointed out that the utilisation of the range and diversity of thermal conditions inside buildings could have significant benefits for comfort and on lowering energy consumption. For example, in a study by Baker and Newsham (1989), a computer simulation model was used to study the effect of occupants' movement on thermal comfort. It was found that, by allowing the occupant to move to the most comfortable position in a room at hourly intervals, for a period during the overheated summer months, overheating (Predicted Percentage Dissatisfied PPD > 20% — PPD is the percentage of time when people are expected to judge ambient thermal conditions as uncomfortable) was reduced from 530 to 115 hours.

Moreover, in assessing the energy implications of having a moving occupant inside an office, Newsham (1990) argued that this might have 'dramatic' effects on comfort predictions, particularly in predictions of overheating. A moving occupant was effectively 0.2°C warmer in under-heated situations and 1.5°C cooler in overheated ones than a fixed occupant. This might suggest that the moving occupant is more adept at avoiding overheating than avoiding underheating. Newsham speculated that if passive overheating control options (movement, shading and ventilation) were all available, then simulation results indicate that it would be best to encourage movement first as it incurs almost no energy penalty on the building.

Similarly, Baker and Standeven (1995) studied spatial and temporal variation in buildings and found that temperatures close to subjects (local temperatures) were 0.5–1.5°C lower than average room temperatures, indicating that the subjects were subconsciously comfort-seeking by making adjustments to position and posture.

The thermal comfort survey

Type and class of thermal comfort survey

The kind of data required and the need to monitor space use patterns of all subjects participating in the survey called for a longitudinal survey, in which a large amount of data is taken from a small number of subjects. Longitudinal surveys thus offer the possibility of linking thermal comfort survey results (i.e. comfort temperatures and prediction equations) to space-use patterns to investigate any effects on occupants' thermal satisfaction.

Such experiments, however, could be quite demanding on subjects (requiring time and effort), which usually result in limiting the number of volunteers together with possible limitations on the number of instruments available.

The thermal comfort survey conducted in this research is a 'class II' field experiment — according to the adaptive model global database classification (de Dear *et al.* 1997) — in which air temperature (T_a), globe temperature (T_g), relative humidity (RH) and air velocity (v) are measured at about the same time as the questionnaires are completed.

Each data set consisted of the following:

- date;
- time;
- thermal comfort vote (C) — 7 categories ranging from +3 'very hot' to –3 'very cold' with 0 as the neutral;
- thermal preference vote (P) — 5 categories ranging from –2 'much cooler' to +2 'much warmer' with 0 as no change required;
- skin moisture (S_w);
- clothing (Clo);
- metabolic rate (Met);
- air temperature (T_a);
- globe temperature (T_g);
- relative humidity (RH);
- air velocity (v).

The data set also included use of fans, cooling or heating, location in the house and ten air temperature and RH readings in selected usable spaces in the house.

All measurements were taken automatically using a data logger housed in a specially tailored case carried by the subject. The various sensors were positioned on top of the case in a way that ensured minimum blockage effects.

Selection of buildings and subjects

The houses selected for the study are located near the centre of Khartoum in two different residential areas of predominantly single-storey houses. Logistical preferences (e.g. ability to arrange contacts and recruit subjects) favoured the selection of four houses with different spatial compositions as

12.1
The four houses selected for the study

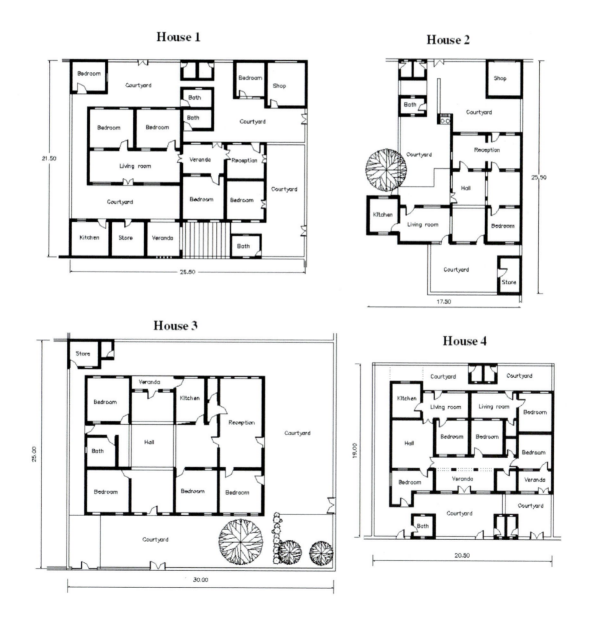

House 1

House 2

House 3

House 4

12.2
Views of house 1

seen in Figures 12.1–12.5. Eleven subjects, familiar with their surroundings and fully acclimatised to living in hot-dry climates, participated in the survey. Their age ranged from 22 to 55 years. Four of them were involved in five experiments, five participated twice and two participated only once.

12.3
Views of house 2

12.4
Views of house 3

12.5
View of house 4

Data analysis

As the questionnaires linked comfort votes to both the time and the place in which they were cast, they provided a starting point for outlining, for each subject, a daily space-use pattern. These patterns were then checked against the findings of the observational study (hourly timetable showing when/where the subjects are). This study was conducted twice. The first time was in conjunction with the thermal comfort survey (2000), while the second was carried out the following year (2001) to further check the accuracy of the produced patterns.

By relating air temperature and RH in different living spaces of the house (between-spaces diversity) to each subject's space-use pattern, the success or failure in avoiding thermal discomfort is assessed. On the other hand, within-spaces diversity is studied by monitoring thermal conditions in ten locations within a certain living space in relation to the space-use pattern inside that space.

Spatial diversity and comfort

Percentage of use versus PPD

If we assume that people — on the whole — are comfort seeking, even at a subconscious level, then a predictable behaviour is to use comfortable spaces as long as possible and to minimise the use of uncomfortable spaces — except for necessary activities and/or when there is no other alternative.

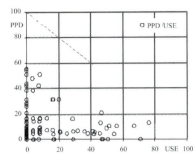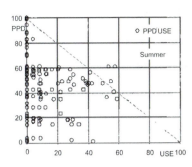

12.6

Percentage of use versus PPD in winter/spring combined (left) and in summer (right)

Therefore, if the percentage of use is plotted against PPD, a hypothetical trend-line (i.e. no factors take precedence over avoiding thermal discomfort) will have maximum use percentages at low PPD levels and vice versa. Figure 12.6 shows PPD/use plot of different living spaces in winter/spring combined and in summer. In winter/spring, the PPD/use pattern followed closely the hypothetical trend-line (dashed line). This is mainly because conditions were generally mild which meant that, at any time, there were comfortable spaces that could be occupied instead of uncomfortable ones.

In summer, however, most of the spaces were uncomfortable during afternoons/evenings, forcing people to use them anyway. The decision then was to occupy the least uncomfortable space. This explains why some spaces were used for about 40% of the time even when their PPD level was as high as 40–50%.

In general, as long as the points of the scatter plot lie within the lower quadrant and close to both axes, this indicates a fairly good utilisation of spatial diversity. For instance, in summer, spaces with high PPD levels were usually used for <20% of the time (concentration of plot points close to the y-axis).

Effect of movement on PPD

PPD for different spaces (space PPD) during each experiment was calculated together with PPD resulting from the subjects moving around the house (movement-PPD). Single-space-PPD is the percentage of time that space is expected to be uncomfortable. Figure 12.7 plots the range offered by different spaces, while the movement-PPD is shown as a green square. The range is not a continuous one, i.e. many single-space-PPD values lie on the line connecting maximum and minimum. The experiments are arranged from winter, spring to summer — left to right. PPD ranges varied between subjects in the same experiment according to sex (men–women segregation) and room allocation restrictions.

As expected, outdoor spaces offered most of the maximum PPD values, with intermediate spaces providing slightly lower values. However, some indoor spaces had over 90% PPD in summer experiments.

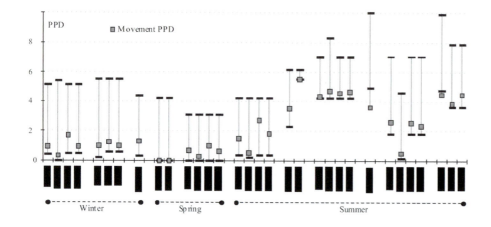

In winter, movement-PPD was generally low, yet some spaces offered a lower PPD for the whole duration of the experiments. In summer, however, movement-PPD was always close to the lowest PPD offered by any single space — in two cases it was even lower than the lowest single-space-PPD. This indicates that, whenever possible, people were moving between living spaces to reduce their overall thermal dissatisfaction (i.e. using the space when it is comfortable and moving to another one when the first becomes uncomfortable).

Utilisation of spatial diversity

People living in traditional courtyard houses often utilise two kinds of spatial diversity: between-spaces and within-spaces diversity. The former describes the difference in environmental conditions between different parts of the house, while the latter describes the diversity inside each living space.

Obviously, the magnitude of between-spaces diversity is far greater than within-spaces as it includes outdoor, intermediate (e.g. verandas) and indoor environments. Monitoring results showed that occupants' movement between different living spaces could potentially effect a 1–2.5 category change in thermal sensation (3–8 K ranges). While the variation within a single living space was quite small (4 K maximum in all houses) which might be expected in high thermal mass buildings.

Between-spaces diversity

Eleven experiments were conducted in the four houses in winter, spring and summer. However, only results from house 1 are presented in this short chapter as an example.

WINTER

Conditions were generally mild and most spaces were quite comfortable for most of the day. Accordingly, Figure 12.8 illustrates the available ranges of

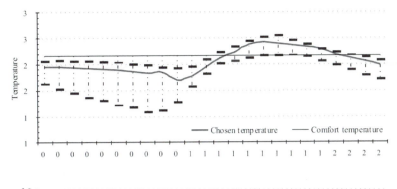

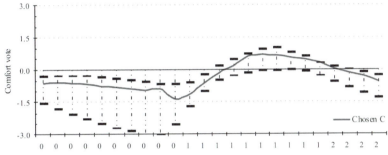

temperature offered by different spaces (ten monitored living spaces) and air temperature chosen for occupation marked as a continuous bold line (obtained as an average weighted by the number of subjects in each location).

To assess whether any departures from the neutral temperature could result in discomfort, the same analysis is also presented in terms of comfort votes. Hourly mean values of comfort were used and the comfort scale was divided into four intervals (1.5/category). The comfort zone thus lies between –1.5 to +1.5.

Using the number of subjects as a weighting factor, total percentages of time were calculated. Subjects in houses 1, 2 and 3 spent 46.9% of their time in comfortable conditions, 20.3% in slightly cool spaces, 30.7% in slightly warm spaces, 1.6% in cool spaces and 0.5% in warm ones. In conclusion, people were quite successful in achieving comfort during winter months.

SPRING

Spring, being a short transition between winter and summer, is characterised by its wide range of temperatures (cold, mild to hot). However, people seem to adapt quickly to these conditions, reflected in their reduced sensitivity to changes in Ta (6°C/comfort scale category). Figure 12.9 shows the wide air temperature ranges that were available and those selected for occupation in house 1. Large departures from the neutral temperature were observed in spring. Diurnal variation in outdoor temperature resulted in 3–6 K departures in selected air temperature from comfort temperature in both directions (hot and cold).

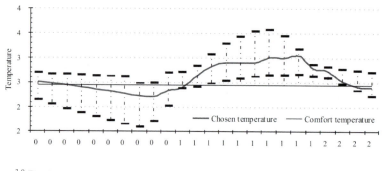

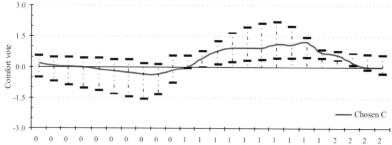

In summary, subjects in houses 1 and 4 spent 3.5% of their time in slightly cool, 69.4% in comfortable and 27.1% in slightly warm conditions. They remained within the comfort zone limits even though the departures from comfort conditions were greater than those in winter. Extreme conditions were successfully avoided most of the time.

12.9
Available springtime ranges of air temperature (top) and comfort (bottom) and those chosen for occupation

SUMMER

Conditions were hot to extremely hot. Minimum temperatures (early morning hours) could be as high as 26–32°C. Three summer experiments were conducted in house 1 and Figure 12.10 is an example that shows temperature and comfort ranges available to subjects with those selected for occupation.

In summary, subjects in the four houses occupied slightly cool spaces for 5.6%, comfortable 27.5%, slightly warm 40.7% and warm to hot spaces for 26.2% of the total time.

It could also be seen that subjects were keen to choose the coolest places to spend their nights, while less emphasis was placed on daytime comfort. In general, unless restrictions were imposed upon space usage, subjects consistently occupied the least uncomfortable spaces available.

Another interesting observation is that subjects preferred to use outdoor spaces with high temperatures and air speeds to indoor spaces with lower temperatures and air speeds. This is seen in the three figures starting at 18:00 hr and continuing throughout the night.

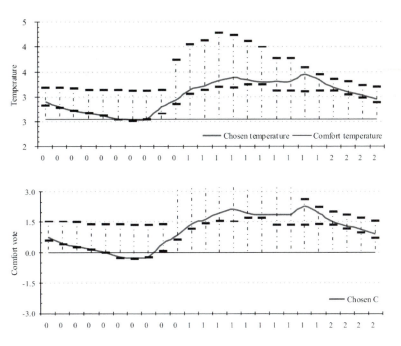

12.10

Available summertime ranges of air temperature (top) and comfort (bottom) and those chosen for occupation

Within-spaces diversity

To investigate the diversity range within indoor living spaces, a long living room in house 1 was monitored for one day in summer as an example. Ten thermometers (A to J) were used as seen in Figure 12.11. To avoid disruption to occupants' daily life pattern, necessary circulation space prevented the positioning of sensors in the middle of the space.

A maximum difference between various locations of 1.6 K is observed between 14:00–16:00 hrs and a minimum of 0.5 K at midnight. The difference is mostly between locations far from and close to windows.

Temporal variation is expectedly higher than spatial, reaching 6.4 K in location E (difference between midday and early morning temperatures). Observing which locations the subjects chose for spending their time revealed that H, G and D were used more than the other locations. Position H — associated with watching TV — was consistently cooler than most locations between 10:00–20:00 hr.

To investigate the utilisation of spatial diversity by the subjects, the difference between local temperatures (recorded close to subjects) and average temperatures (taken at the centre of the room in most cases) is calculated. Theoretically, a good utilisation would suggest that local temperatures are expected to be higher than average ones in cold conditions and lower in hot ones. Table 12.1 lists the average difference in the three seasons. Means for data above and below comfort temperatures were calculated separately to investigate any significant difference in behaviour.

In winter, 60.9% of the time local temperatures were lower than average temperatures suggesting that subjects were not successful in

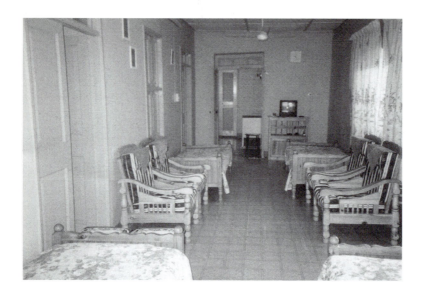

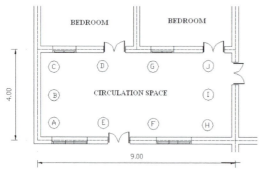

12.11
A view of the living room taken from location (B) towards (I) — (top) and the distribution of sensors across the living room (left)

utilising within-spaces diversity. Spring is quite difficult to judge, as the objective changes throughout the season. On cold days, higher local temperatures are advantageous and vice versa on hot days. Nonetheless, the mean local temperature was 0.14 K higher than the average one.

Table 12.1 **Difference between local and average temperatures in each season**

	Mean diff.	Max diff.	Higher	Lower	Equal
Winter	−0.30	2.70 K	35.2%	60.9%	3.9%
below T_c	−0.43		33.3%	63.5%	3.2%
above T_c	−0.20		36.6%	58.8%	4.6%
Spring	0.14	2.85 K	57.1%	40.4%	2.5%
below T_c	0.04		50.0%	47.8%	2.2%
above T_c	0.17		59.8%	37.6%	2.6%
Summer	−0.11	3.95 K	44.2%	52.8%	3.0%
below T_c	−0.36		27.5%	70.0%	2.5%
above T_c	−0.10		44.8%	52.2%	3.0%

In summer, mean local temperature was 0.11 K lower than the average, indicating a fairly good utilisation of within-spaces diversity. Local temperatures were lower in 52.8% of the time and higher in 44.2%.

In summary, the range of within-spaces diversity is quite small (4 K maximum) which might be expected in traditional buildings where temperature variation inside rooms tends to be small except when large glazed surfaces are used. However, if the temporal variation is taken into account the difference is far greater as some locations inside a room, for instance, will fluctuate more freely than others (e.g. near openings).

Synthesis and conclusions

The previous discussion and analysis highlighted many interrelated topics. Of these, we stress only three:

a   the energy saving potential of spatial diversity in residential buildings;
b   house design decisions (spatial composition); and
c   some important factors that take precedence over thermal comfort in some situations.

Energy saving potential of spatial diversity in housing

A simple observation illustrating the usefulness of having spaces varied in character and spatial composition is night-time comfort in summer months. Monitoring results showed that by sleeping in the courtyard (under the clear sky) people in traditional courtyard houses had comfortable sleep throughout the summer — Figure 12.12 (Merghani 2001).

Sleeping in air-conditioned indoor spaces or in naturally-cooled outdoor ones makes a big difference in energy consumption. The potential energy saving could be considerable in magnitude bearing in mind that Khartoum is hot for almost nine months of the year. To illustrate this, Figure 12.13 shows peak electricity loads on the Sudanese national grid for 10 May

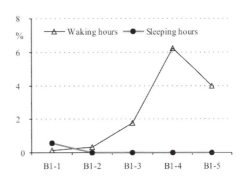

12.12
Percentage of waking and sleeping hours spent in uncomfortable conditions

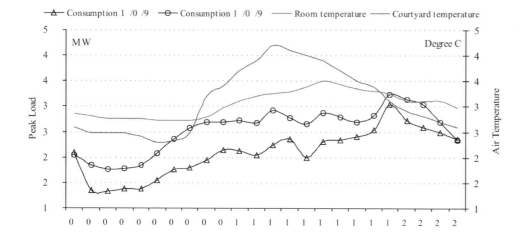

1998 and 1999. Plotted on the same chart is air temperature inside a naturally-ventilated room and a courtyard for the same day in 1999.

The peak load occurs between 17:00–21:00 hr. As working days end at 14:00 hr (drop in consumption lines), air conditioning and lighting in domestic buildings may account for most of the increase in energy consumption after work hours. This, obviously, highlights the great potential for energy saving if air-conditioning use during evenings and early night hours is reduced by using naturally-cooled outdoor living spaces or courtyards (comfort temperature in summer is 30.9°C).

12.13
Peak electricity load on the Sudanese national grid (National Electricity Corporation, 2000) — plotted on the same chart are the air temperatures inside a naturally-ventilated room and a courtyard for the same day

Implications for house design

A graded set of living spaces offering varying degrees of enclosure provides the occupants with thermal environments that can be used at different times of the day in different seasons.

From the findings regarding between-space and within-space variability one could speculate a hypothetical utilisation of both spatial and temporal variation. For example, a certain house design provides, say, four living spaces, each space being comfortable for only 6–8 hours during the day (i.e. each space has up to 66–75% PPD level) at a different time from the other three. Accordingly, a subject free and willing to move between these spaces could achieve a movement-PPD of almost zero if he/she seeks out the most comfortable location. This is outlined in Table 12.2 below, taking a typical summer day as an example. Living spaces could be:

L1 = a courtyard or a roof terrace;
L2 = a west side living space or a veranda;
L3 = a basement or a living room with an evaporative cooling device; and
L4 = a courtyard, veranda or a sleeping/roof terrace.

Table 12.2 **Space-use patterns and the utilisation of spatial and temporal variation (grey-coloured cells are uncomfortable hours while black-bordered ones are those selected for occupation)**

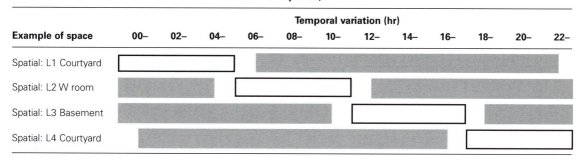

Factors limiting free choice of living spaces

As we have seen, people were making fairly good utilisation of the available range of conditions between and within living spaces. However, a multitude and a variety of reasons could limit occupants' free choice of space such as:

- functional reasons, e.g. using an uncomfortably hot kitchen for preparing food or not using it when it is comfortable because of lack/existence of special-use furniture;
- restrictions being imposed upon the use of certain spaces, e.g. men/women segregation, guest rooms kept unused for tidiness, privacy, etc.;
- people continue to use their allocated rooms even when they are uncomfortable. They may also use certain spaces because they feel attached to them;
- people use spaces to watch TV, chat and enjoy the company of others;
- economic factors, e.g. being unable to afford the initial and running costs of air-conditioning.

Figure 12.14 is a breakdown of the important factors that took precedence over thermal comfort based on reasons offered by the survey

12.14
The importance of some factors that took precedence over thermal comfort during the thermal comfort survey

35% — Prefer outdoors
22% — Men/women segregation
16% — Working
15% — Watching TV
7% — Chatting/socialising
5% — Using allocated room

subjects themselves. The list is not exhaustive as the reasons are quite subjective.

Concluding remark

It has been shown that, if given the chance, people will make the best of the spatial diversity available. This good utilisation, however, is part of a complex (conscious and subconscious) processes through which people create their own criteria to evaluate the acceptability of the environments they choose to occupy. The ongoing research into occupants' satisfaction and thermal comfort inside buildings has been, mostly, quantitative in nature. The work suggests that it is important to develop the notion of appropriate diversity and range rather than seeking further precision.

References

Al-Azzawi, S. (1996a) 'Seasonal impact of climate on the pattern of urban family life; indigenous courtyard houses of Baghdad', in Sayigh, A. A. M. (ed.) *Proceedings of the World Renewable Energy Congress*, Oxford: Pergamon Press.

Al-Azzawi, S. (1996b) 'Daily impact of climate on the pattern of urban family life; indigenous courtyard houses of Baghdad', in Sayigh, A. A. M. (ed.) *Proceedings of the World Renewable Energy Congress*, Oxford: Pergamon Press.

Baker, N. V. (2000) 'We are all outdoor animals', in *Proceedings of PLEA 2000: Architecture-City-Environment*, Steemers, K. and Yannas, S. (eds), London: James & James.

Baker, N. V. and Newsham, G. (1989) 'Comfort studies by spatial modelling of a direct gain room', in *Proceedings of the 2nd European Conference on Architecture*, Paris, Kluwer.

Baker, N. V. and Standeven, M. A. (1995) 'A behavioural approach to thermal comfort assessment in naturally ventilated buildings', in *Proceedings of CIBSE National Conference*, Eastbourne, UK.

de Dear, R., Brager, G. and Cooper, D. (1997) *Developing an adaptive model of thermal comfort and preference — Final Report*, RP-884: ASHRAE.

Fitch, J. M. (1972) *American building; the environmental forces that shape it*, 2nd edn, Boston, Mass.: Houghton Miffin Co.

Heerwagen, J. H., Loveland, J. and Diamond, R. (1991) *Energy edge: Post occupancy evaluation project, Final Report* (USDOE Contract No DE-AC06-89RLL11659, BPA Contract No. DE-B179-90BP04252), Seattle: University of Washington.

Heidari, S. (2000) 'Thermal comfort in Iranian courtyard housing', unpublished PhD thesis, University of Sheffield.

Heschong, L. (1979) *Thermal delight in architecture*, Cambridge, Mass.: MIT Press.

Humphreys, M. A. (1997) 'An adaptive approach to thermal comfort criteria', in Clements-Croome, D. (ed.), *Naturally ventilated buildings; buildings for the senses, the economy and society*, London: E & FN Spon.

Knowles, R. (1999) Rituals of space (Online) available: http:www-rcf.usc.edu/~rknowles/rituals_place/rituals_place.html (accessed 25 May 2004).

Markus, T. A. and Morris, E. N. (1980) *Buildings, climate and energy*, London: Pitman Publishing Limited.

Merghani, A. (2001) 'Thermal comfort and spatial variability: a study of traditional courtyard houses in the hot dry climate of Khartoum, Sudan', unpublished PhD thesis, University of Cambridge.

Newsham, G. R. (1990) 'Investigating the role of thermal comfort in the assessment of building energy performance using a spatial thermal model', unpublished PhD thesis, University of Cambridge.

Roaf, S. (1988) 'The windcatchers of Yazd', unpublished PhD thesis, Oxford Brookes University.

Rodger, A. (1974) 'The Sudanese heat trap', *The Ecologist*, 4(3): 102–6.

PART 5

Design

Chapter 13

Experiencing climate: architecture and environmental diversity

Peter Fisher

Introduction

> Architecture gives a material structure to societal institutions and
> to daily life, reifying the course of the sun and the cycle of the
> hours of the day.
>
> (Pallasmaa 1996)

In this brief but eloquent summary of architecture's capacity to order daily life Pallasmaa implies that the tempering of climate ought to be one of the most poignant and poetic generators of architectural form. The idea that environmental moderation can shape and refine architectural ambitions is, of course, not new. As Rapoport (1969) and Oliver (1997) have illustrated, environmental responses inspired by climatic factors have had the potential to locate or 'ground' vernacular architecture across the world to a specific region, and at a more local level, to articulate use, render legible the passage of time and qualify movement through space. But what role has environmental moderation played in modern architectural design?

The development of mechanical environmental systems during the twentieth century has denied to much recent architecture the symbolic role that it previously played as a form of 'shelter' which by necessity responded to the prevailing local climate. It is now possible, at the expense of consuming greater levels of energy, for a building's form and orientation to ignore climate completely. Not only has this had a severe effect on emissions but it has also suppressed many of the experiential qualities that a concern with climatic factors had previously inspired, and can be seen as part of a general progression towards an increasingly homogeneous world in which cities and buildings throughout the globe have begun to look the same, irrespective of location. Site-specific relationships between the ambient climate and orientation, use and movement are now frequently the exception rather than the rule.

Earlier in the twentieth century, however, this was not always the case. Having grown out of the increased availability of industrial technology and widespread concern over the environment of industrial cities in the late nineteenth century, it was only by the middle of the twentieth century that mechanical systems were sufficiently advanced to allow the complete divorcing of a building's interior from its ambient climate. Importantly, though the first fully air-conditioned inhabited building was a 1928 office block in Texas (Banham 1969), it was not until after the Second World War that advanced mechanical environmental systems were commonly adopted. Consequently, despite a concern to articulate new attitudes to both spatial design and construction, most first generation modernism was built before the widespread adoption of mechanical systems.

The idea that an architectural agenda can be enriched by climate's inclusion is usefully illustrated by a number of early and mid-century modernist examples. It is particularly useful to look at buildings completed both before and after the widespread adoption of the mechanical system. The primary lessons contained in the examples discussed here lay in the spatial response to clear climatic differences in the two pre-war buildings and in the attitude to technology and the mechanical system in the two post-war buildings. What the examples have in common is that, while they employ varying levels of environmental technology in various locations and at different times, all make the encountering and tempering of climatic variation a conscious, tangible and legible architectural aim at a fundamental conceptual level. In each case the impact of this thinking on the overall design is worth discussing at some length in order to demonstrate how an embrace of climatic diversity can inform the process of developing and resolving design proposals.

Le Corbusier and Kahn: post mechanical system and the attitude to technology

Whereas the responses of the pre-war buildings, discussed in detail later in this chapter, to climatic diversity were partly pragmatic and a reflection of the available technologies, the two post-war buildings in question, Le Corbusier's Mill Owners' Building in Ahmedabad (1954) and Kahn's Mellon Centre at

Yale (1972), provide useful examples of different post-war attitudes to technology. They have in common a conscious and explicit, but very distinct, response to the mechanical system that contrasts with the prevailing attitude of the time. As such they are interesting not only because they exhibit quite different nuanced responses to climate, but also because they embrace climatic variation and the diversity it brings while exploiting vastly differing levels of environmental technology.

Le Corbusier's rediscovery of the building form as an elemental environmental filter can be contrasted with Kahn's highly articulate integration of mechanical systems into a building whose interior environment is subtly attuned to and qualified by the variation in natural light levels to which it is exposed, but in a building that is largely mechanically conditioned. The essential difference between the two being that in the Mill Owners' Building the response of the building fabric to environmental diversity is the primary means of environmental control as well as being experiential. In the Mellon Centre the primary means of environmental control is mechanical and the climatic response is mostly experiential. As such, they demonstrate how the adoption or not of available and appropriate technology does not necessarily preclude a sensitivity to environmental diversity and, importantly, to its experience, even in buildings where a highly controlled environment is required. And, together, the two demonstrate that environmental control can be an important experiential component, irrespective of whether or not it is the primary method of climatic moderation.

It is worth noting at this point that for both architects the role technology should play in determining the character of interior environments in the light of climatic variation was constantly in question. Rather than making a once-and-for-all judgement on the issue they reconsidered the appropriate approach for each and every project. Consequently, their respective positions on the issue would be reversed if the buildings being discussed were, say, Le Corbusier's Carpenter Centre and Kahn's Ahmedabad Management School.

The Mill Owners' Building: the facade as climatic filter

In the Mill Owners' Building (Figure 13.1) the rejection of the mechanical system, the articulation of the brise-soleil and the permeable membrane, with their collective response to climate are the key environmental moves within the building.

At the Mill Owners' Building brise-soleil dominate the design of both the east and west facades. Essentially an exclusive club with philanthropic pretensions, the building's environmental strategy needed to ensure that an appropriate environment for business meetings, receptions and lectures was provided, and thus the prevention of solar gain was a major concern. Ahmedabad, which has no heating degree days at all, has a relatively 'cool' and dry winter, during which the average temperature is marginally

13.1
Le Corbusier's Mill Owners' Building in Ahmedabad (1954)

above 20°C. A warm and wet summer from June to August is preceded by what is in fact the hottest period of the year from April to June, when peak temperatures reach 40–45°C. The prevailing wind direction in the warmest period is from the south-west.

Le Corbusier's rediscovery of the facade as a climatic filter was not solely inspired by a concern for environmental strategy, but was as much a wider criticism of the universality of modernism. Frampton has referred to this gradual evolution in approach as 'the monumentalisation of the vernacular' (Frampton 1992). The gradual development of an attitude to the facade pre-war had been to reduce it to a hermetically sealed continuous membrane and then to rely on a complex mechanical system to attain a habitable internal climate. However, as is well known, due to various factors particularly in the case of the Paris Refuge, this approach was little short of a disaster. Indeed, perhaps because of the problems suffered in his earlier work, the facade subsequently became an important component in Le Corbusier's search for a more rooted, material and experiential architecture in which the envelope became a technically low-key, primal climatic filter in which simple elements could fulfil quite complex environmental roles.

Le Corbusier's reawakened interest in the climatic role of the facade following the Second World War can be separated into two distinct phases. First, he invents the fixed brise-soleil which is applied to the exterior of the membrane, altering the conditions to which it is subjected. Second, he rethinks the actual membrane itself which becomes simpler and more permeable to the external climate. The former is concerned with an inherent response and the latter an interactive response to climatic moderation.

The brise-soleil may indeed be a strong sculptural device, but it is clear from various sources that it was actually inspired by the techniques used for shading traditional vernacular buildings that Le Corbusier had observed in Algeria. As Colquhoun has noted, the brise-soleil became Le Corbusier's post-war signature, much as the use of piloti had dominated his earlier work. It was symbolic of a more specific and regionally responsive architecture, although ironically he adopted it on an almost universal basis. Colquhoun sums up the complexity of the role played by the brise-soleil as follows:

> The ideal transparency of the external wall was not abandoned; its effects were counteracted by the addition of a new tectonic element. But the brise-soleil was more than a technical device; it introduced a new architectural element in the form of a thick, permeable wall, whose depth and subdivisions gave the facade the modelling and aedicular expression that had been lost with the suppression of the window and the pilaster.
>
> (Colquhoun 1991: 107)

While the north and south side walls are completely blank and therefore closed to the sun, the east and west facades are much more open, and face the street and the neighbouring Sabarmati River respectively. To the west, the 45° angle of the brise-soleil protects against afternoon sun all year round and helps the facade to capture the prevailing south-west winds (only in winter are small amounts of sunlight allowed in after 4.00 p.m.). It also creates a slightly more closed, more honorific entrance facade. An angle of 45° seems strikingly formal but it happens to correspond exactly to the prevailing wind direction in the warmest months. In contrast, the east facing brise-soleil is more shallow, with vertical fins that are perpendicular to the building, allowing a less restricted view to the external landscape. The sun-path diagram for this latitude indicates that this geometry allows sunlight penetration until 9.00 a.m. in all seasons, an aspect of the building's environmental strategy that can only be climatically justified when the nights are cool and the building needs warming up in the morning.

The primary environmental justification of the facade geometry is the passive control of climate through built form. As shading and wind-channelling devices, the brise-soleil are quite effective, mitigating solar gain at the hottest times of the day and encouraging useful cross-ventilation. What is more, they frame and direct views of the surroundings and give depth and relief to the facades. The design is not, however, without its drawbacks. One recurrent and pertinent criticism of Corbusier's Indian work is that in this climate a thermally massive concrete brise-soleil will heat up under the intense summer sun and effectively pre-heat incoming night-time breezes, hindering the building's ability to cool down overnight.

In the Mill Owners' Building, Le Corbusier's rejection of the mechanical system was clearly part of a conscious reaction and a return to more primal and elemental qualities. Importantly, the means of passive environmental control is 'monumentalised', transforming it into something, elemental, spatial and experiential. The role and subsequent experience of the facades is well beyond that of a simple technical device, but has its roots in a pragmatic response.

13.2
Kahn's Mellon Centre at Yale (1972)

The Mellon Centre: climatic response and the integration of technology

In the Mellon Centre (Figure 13.2) air-conditioning was considered a programmatic necessity and was accepted and legibly exposed as such by Kahn. Yet, despite the acceptance of the mechanical system, climatic variation is subtly used to articulate and qualify movement and experience within the building.

New Haven's climate, in the north-eastern USA, is quite harsh, with warm, humid summers and very cold winters. It was concern for conservation and the precise control of heat and humidity that required the use of air-conditioning. The ducts of the mechanical system are exposed on the lower floors and contained in the V-shaped concrete beams on the top floor. Clearly, legible as air-conditioning, but carefully and beautifully detailed nonetheless, there is never any doubt that the climate is being largely mechanically modified.

The Centre is located on a tight urban site in the centre of New Haven, close to the University and to the earlier Yale University Gallery, also by Kahn. The facades of the main elevation are built up to the street edge, with shops maintaining the street's integrity at ground level. The mostly closed upper elevations, clad in steel panels and glass, are quite reticent, with controlled views in and out. The disposition of glazing subtly indicates the location of important functions, such as the library and print room.

Upon entering the building, the reticence of the facades gives way to a top-lit, daylight filled entrance court, the first of two such courts. The building houses both an art gallery for the display of British Art and a library, rare prints room and study areas. These are arranged around the two courts, giving an entrance court and an academic court, with different qualities. Kahn and Paul Mellon, the benefactor, envisaged a building with the subtle experiential qualities of the English country house, from which most of the works would have originated. As Hawkes has pointed out, for Kahn, the challenge was to integrate the technologies of his own time within architecture that attempted to recreate the environmental and spatial qualities of an earlier building tradition:

> The entire building is covered by a rooflight system of the utmost simplicity. Each structural bay of the building is surmounted by V-shaped concrete beams which provide space for mechanical services and become splayed reveals to the rooflights. Each bay has four simple dome lights which in the picture galleries have external metal louvres to exclude direct sunlight. Beneath are opal diffusers which further control the quality and quantity of light, rendering it, in its uniformity, closer to the relative flatness of English light under which these paintings were produced and which they frequently depict.
>
> (Hawkes 1996: 205)

At the upper levels, both courts are part of the circulation within the gallery spaces. The tall entrance court contains no pictures and therefore the rooflights in this area require no solar shading, whereas the academic court is used for display and has shading above the glazing. While both courts are structurally the same — though the academic one is three rather than four storeys high — there is a subtle distinction between them. The light in the unshaded entrance is more dynamic, which seems appropriate for movement and arrival, whereas in the shaded academic court the light is more subdued and reflective. In what is a relatively introvert building, the two courts provide the main points of visual reference and orientation. Views into the courts from the surrounding spaces allow the dynamic quality of continuously changing patterns of sunlight and shadow to be perceived throughout the building. The contrast between the warmer, more dynamic lighting of the timber-lined courtyards and the cooler, more even quality of light provided within the galleries helps to pace and order movement and to articulate the key spatial system.

As well as the borrowed top-light from the courts, the galleries are also side-lit, giving a range of lighting conditions for the pictures displayed. The visual contrast helps the galleries to seem calm and thus encourages concentration on the pictures and their fine level of detail, while at the same time orienting visitors within the building and providing a degree of visual interest that helps to counter 'museum fatigue'. The windows in the display areas, like the offices, print room and library as well, have beautifully detailed, sliding and tilting timber shutters internally, which allow localised control of light. In doing so they give a very intimate and personal level of interaction within what is a large and mostly controlled building. Added to which, the windows further localise the experience of the building by framing key views out of the building, linking the galleries to the wider context of New Haven and Yale University.

At the Mellon Centre, Kahn succeeds in providing spatial and temporal environmental diversity through an intelligent synthesis of spatial ideas and intuitions about the character of the interior climate he wished to create. Though divorced from the ambient thermal climate, visitors are always aware of changes in the character of natural light that is available, and an appropriate and stimulating setting for the display and study of paintings is the result.

The obvious and critical difference between the Mill Owners' Building and the Mellon Centre, is in the philosophical attitude to the mechanical system. But in both buildings the means of environmental control is clear and the legible. In one it is mostly spatial and the other mainly mechanical. But again in both, as in Viipuri and Como, the overall form and key spatial sequences are informed by climate and both are richer buildings for the inclusion of environmental diversity. In both that inclusion of climatic experience also roots the buildings in their localities, irrespective of their relative technical sophistication.

Aalto and Terragni: pre-mechanical system and the spatial stance

Aalto's Viipuri Library (1935) and Terragni's Como Kindergarten (1937) are revealing because at each latitude the climate dictates a clear set of environmental concerns. They were completed before an easy resort to technology was possible and at a time when universal internationalism prevailed. Both can be described as belonging within the then prevalent international style, yet both exploit climatic typologies that are highly appropriate to their specific context, and which, consciously or not, can be traced to local traditions. In each case the handling of the predominant climatic concerns allows environmental strategy to become part of the building's defining poetic.

Viipuri in Finland (now Vyborg in Russia) and Como in Italy are at the geographical and climatic ends of Europe (at latitudes 61° north and 44° north respectively). The principal environmental task of a building in Finland is to provide heating for almost the entire year and to respond to the low levels of daylight during the winter, whereas in northern Italy it is to provide some heating during winter and protection from sun during summer. In the library the principal concern is daylight, while in the kindergarten it is sunlight, resulting in enclosure and protection from a harsh, dark climate on the one hand and transitional filters that both allow and prevent the ingress of sunlight at different times of the year on the other.

Viipuri Library: the stable enclosure

At Viipuri, the library is separated into two volumetric elements: the reading room and the ancillary functions (lecture room, offices, WCs, etc.) which express and render legible the main function of the building. The reading room is the differentiated space that orientates the entire building; the main function onto which all other functions are hung (Figure 13.3). Essentially an atrium, it was a typology present in all Aalto's libraries, culminating in his masterpiece at Seinäjoki.

Not only is the reading room a distinctly separate volume, it is also radically different to other spaces within the building in conceptual terms. In a 1926 article Aalto wrote that the sharp differentiation between the warm interior and the surroundings dictated by the Finnish climate was not in itself wrong, but that the ceremonial link back to outdoors was often immature. This intuition was summarised as follows:

13.3
Aalto's Viipuri Library (1935)

Our cold climate might do violence to the unity which should link the interior and exterior of our homes, with the result that the entrance section cannot be given the elegant ceremonial form it has in the civilised climatic zones of the south. The fault, however, is hardly the climate's, it is more likely due to immature form. There is nothing wrong with our homes being closed to the outside world — so are those in the south, though for different reasons — but the screening element of our houses is almost invariably placed badly.

(Schildt 1984)

He went on to extol the virtues of the English hall and the fact that it 'symbolised' the open air under the home roof. The essence of Viipuri's spatial conception is apparent in the article, which Aalto tentatively begins to formulate as an expression of the relationship between architectural form and climatic context.

The reading room can be interpreted as a metaphorical reinterpretation of outdoor space, which is thus climatically more stable. It is quite widely acknowledged that Asplund's Skandia cinema of 1923, where the ceiling is a painted Mediterranean sky, was a major influence on Aalto (Schildt 1984; Wrede 1980). Richard Weston has written that instead of the recreated southern sky Viipuri 'materialises the atmospheric northern sky as a low, misty veil hovering above the encircling horizon of books' (Weston 1995). The quality of light is indeed external and overcast, a quality maintained even without sufficient natural light.

The building consists of essentially three different types of space, the conventional 'rooms' of the ancillary space which act as a backdrop, the metaphorically 'outdoor' reading room and the entrance sequence which introduces the reading room and is important in establishing the link and transition from outdoors to in.

The spatial order of the library is determined by the location and character of the reading room and of the spatial sequence that articulates the transition from the natural world outside to its metaphoric recreation inside. The visitor is gradually taken into the interior, through the foyer entrance hall with its distanced contact to outside, then through a low dark space beneath the reading room before rising into the well lit 'outdoor' reading room. The foyer is naturally lit via the staircase to the first floor offices, a screen through which the visitor is still aware of the world outside. Whereas the reading room is entirely about diffuse light, the staircase does allow filtered sunlight into the foyer. Facing east, the space receives whatever morning sun is available. More important than the actual transparency of a view is the implied transparency of a series of filigree screens through which light is dispersed and diffused.

The reading room actually has two functions — it is a reference library and a lending library. But Aalto wanted to create one main space at the centre of the building, which was spatially distinguished from the secondary functions. Bringing the two functions together enabled him to make one, single volume that is articulated by the change in levels and the stepped ceiling. Depending on the amount of natural light available, environmental conditions in the reading room fluctuate between the natural and the artificial (though they are always metaphorically 'natural'). It is lit by 57 cylindrical rooflights, which are of sufficient depth to ensure that the highest summer sun (at a solar altitude of 52°) never enters the space directly. The natural light in the reading room is diffuse and directionless and, like the overcast sky, the light source is never seen directly. The light is spread evenly throughout the whole space, so that even a reader standing in front of the shelving casts no significant shadows onto the books. Due to the location of the light sources high up in the ceiling the glare often associated with side lighting is not present. The rooflights exploit the fact that an overcast sky is three times as bright at the zenith as at the horizon, enabling less glazing for the same amount of light. In a climate where during December the average external luminance is 640 lux and the average temperature is −3.9°C, balancing the amount of glazing relative to daylight ingress and heat loss is important.

Como Kindergarten: the transitional filter

The kindergarten (Figure 13.4) is orientated around an open garden court, which, as with the reading room at Viipuri, orientates the whole building. But though a physically external space (with a real sky), the court is read as a room within the building. The spatial ambiguity of the court begins to hint at the conception of the whole building. At a theoretical level it is consistent with both Aalto's and Le Corbusier's position regarding the need to break down barriers between inside and out — with the implementation tending towards Le Corbusier's linking of internal and external space rather than Aalto's enclosure and recreation of external space.

It is the transitional 'filters' between inside and out that help to fine-tune the otherwise relatively simple organisation of space. They are highly articulated, treated not merely as a series of two-dimensional layers but as three-dimensional volumes which help to elaborate key spatial relationships and encourage particular patterns of activity. The connection between inside and out is extremely strong. Terragni's admiration for contemporary Dutch architecture, in particular the work of Duiker, is well known (Schumacher 1980) and it is not difficult to see a philosophical link with the sheltered external play areas of Duiker's earlier Open Air School (1930).

In the kindergarten, Terragni 'used a discrete separation of structure and enclosure in special locations to accentuate the theme of shift' (Schumacher 1991), which creates spatial ambiguity. One can feel physically inside but spatially outside, or vice versa. This allowed Terragni to indulge in his favourite formal device, diagonally shifted rectangles, to create

**Giuseppe Terragni's Como Nursery
School (1937)**

spatially ambiguous volumes. It is difficult to identify the precise point at
which one space stops and another begins. The spatially defining elements
sometimes stand outside the physically defining elements and sometimes
inside.

The kindergarten consists essentially of a roof which is a
continuous membrane stretched across the building and under which are
placed a range of transitional filters that are able to respond to orientation.
The notion of a building having differently configured facades according to
orientation is something Terragni wrote about in his report on the other major
building he designed in this period, the Como Casa del Fascio (Terragni 1936).
In the section entitled 'Orientation and Tradition', he gave an account of his
use of Neufert sunpath diagrams and the benefits of Como's solar
orientation, writing that 'the Casa del Fascio ... [mirrors] in its four facades
the fundamental concept of the four different insulation and illumination
conditions' (Schumacher 1991). It is an idea carried through into the
kindergarten. The roof is a neutral, static barrier, whereas the filters are

responsive: opening to the south, closing to the north and providing protected play areas adjacent to the classrooms.

The way that the configuration of filters varies according to orientation is probably at its most apparent within the court. The north-facing classroom corridor is largely opaque, being only one-quarter glazed with the structure penetrating the wall and standing in the court. The long band of glazing running the full length of the wall, however, ensures that the contact to outside is maintained, but its more enclosed nature suggests that the orientation of the wing is in the opposite direction. The facade of the south-facing dining room is the complete opposite, being entirely glazed and thus capturing the winter sun. Here, the actual facade is pulled away from the main structural line out into the court and returns to allow an entrance to be formed. Moveable canvas blinds sit within the structural facade which, when pulled down, create a narrow space that is literally between inside and out. It is interesting that Terragni exploited the need for diversely configured facades at different orientations to enliven the building's form. They are probably the elements that Zevi was thinking of when he wrote of

> the poetic stance which ... enlivens ... the corners and junctions, the surfaces and volumes, the links with the ground, the transparencies and connections to the sky.
>
> (Zevi 1989)

Two of the most eloquent and joyous filters are those of the entrance and the classrooms. In both, the main shading element is pulled away from the facade to fulfil a role beyond its function. At the entrance on the western elevation, the structural support penetrates the glazing to form a frame containing an external canvas blind that both articulates and shades the entrance balcony. The filter along the southern elevation of the classroom block is the most poetic shading device of all. The structurally independent frame in front of the facade defines the spaces even when closed, but when open it makes sheltered play areas outside between the classrooms and the garden. At first glance it is possible to assume that the external blinds provide rather poor shading to the classrooms. In fact they simply shade the external play areas and the principal protection for the classrooms is the deep overhang, backed up by internal blinds. The midday sun is completely excluded during June and only begins to penetrate significantly during March and September.

In both Viipuri and Como a nuanced, spatial engagement with the local environment is immediately apparent. Design decisions concerning overall form and the more detailed spatial arrangement of walls, roofs and apertures express and elaborate environmental strategies that respond to local climatic conditions. Both engage the environment as part of a subtle holistic game in which decisions about the overall form and key spatial sequences are informed by a desire to exploit the positive benefits, or to counter the most obvious drawbacks of the local climate. The library thus encloses and protects while maximising the ingress of daylight and the nursery school alternately allows and prevents the ingress of sunlight.

Conclusion

Conventional architectural theory rarely considers environmental moderation to be a critical conceptual component of architecture. The buildings discussed here employ varying levels of environmental technology in various locations, yet all make the encountering and tempering of climatic variation a conscious, tangible and legible architectural aim at a fundamental conceptual level. Their articulation, elaboration and structuring of human experience as shelter makes them richer and more enjoyable places to inhabit, while their exploitation of, wherever possible, passive environmental control also gives them more legitimacy regarding energy consumption.

It is worth noting that much current 'green' architecture seems to have been reduced to a form of carbon accountancy (despite the fact that many of the most lauded examples fail even in that regard when properly examined). The common assumption that environmental strategy is a matter of detailing or of making late modifications to designs dictated by other parameters has in many cases meant that the envelope has been reduced to little more than an element to 'prevent people falling out of the building' (Banham 1969). As technology improves, and the production of the entire energy needs of the building on site becomes feasible, it is clear that buildings producing zero emissions will become more than possible; indeed probably the norm. They will contribute, however, to the further impoverishment of the built environment if the spatial and experiential issues that need to be addressed in envelope design continue to be ignored.

The recent general growth in ecological concerns has helped to refocus attention on the idea that environmental moderation remains one of architecture's primary tasks. In the drive to reframe appropriate architectural ambitions, the question of how to make 'local' architecture is being reassessed from a number of perspectives. Most obvious are the material 'local' demands facing architects. Yet it cannot be overemphasised that the extent to which a building embraces its local environment, and thus encourages interaction with and awareness of that environment, may be just as significant to its 'local' character. Allowing a building to embrace rather than divorce its climatic context has the power to enrich and animate architectural experience because it engenders environmental diversity that binds architecture to its locality in experiential terms.

Two final thoughts are prompted by this study. First, how the careful orchestration of the dynamic relationships between interior and exterior environments in the buildings reviewed here leads to a close marriage of spatial and environmental strategy. Indeed spatial, environmental and landscape strategy are in an important sense shown to be one and the same thing. Second, it is clear that the more widespread adoption of climatically-responsive thinking by contemporary designers ought to have the welcome effect of prompting a more diverse range of climate-specific environmental strategies and thus site-specific buildings.

References

Banham, P. R. (1969) *The Architecture of the Well-tempered Environment*, Chicago: University of Chicago Press.

Colquhoun, A. (1991) *Modernity and the Classical Tradition*, Cambridge Mass.: MIT Press.

Frampton, K. (1992) *Modern Architecture: A Critical History*, 3rd edn, London: Thames and Hudson.

Hawkes, D. (1996) *The Environmental Tradition*, London: Spon.

Pallasmaa, J. (1996) *The Eyes of the Skin: Architecture and the Senses*, London: Academy Editions.

Oliver, P. (ed.) (1997) *Encyclopedia of Vernacular Architecture of the World*, Cambridge: Cambridge University Press.

Rapoport, A. (1969) *House Form and Culture*, Englewood Cliffs, N.J: Prentice-Hall.

Schildt, G. (1984) *Alvar Aalto: The Early Years*, New York: Rizzoli.

Schumacher, T. (1980) *The Danteum*, London: Triangle Architectural Publishing.

Schumacher, T. (1991) *Surface and Symbol*, New York: Princeton Architectural Press.

Terragni, G. (1936) Relazione tecniche, *Quadrante*, October 35(36): 38–55.

Weston, R. (1995) *Alvar Aalto*, London: Phaidon Press.

Wrede, S. (1980) *The Architecture of Erik Gunnar Asplund*, Cambridge Mass.: MIT Press.

Zevi B. (1989) *Giuseppe Terragni*, London: Triangle Architectural Publishing.

Index